# MANAGING DIGITAL RISKS

## A PRIMER

DECEMBER 2023

ADB

ASIAN DEVELOPMENT BANK

© 2023 Asian Development Bank
6 ADB Avenue, Mandaluyong City, 1550 Metro Manila, Philippines
Tel +63 2 8632 4444; Fax +63 2 8636 2444
www.adb.org

Some rights reserved. Published in 2023.

ISBN 978-92-9270-557-2 (print); 978-92-9270-558-9 (electronic); 978-92-9270-559-6 (ebook)
Publication Stock No. TIM230604-2
DOI: http://dx.doi.org/10.22617/TIM230604-2

The views expressed in this publication are those of the authors and do not necessarily reflect the views and policies of the Asian Development Bank (ADB) or its Board of Governors or the governments they represent.

ADB does not guarantee the accuracy of the data included in this publication and accepts no responsibility for any consequence of their use. The mention of specific companies or products of manufacturers does not imply that they are endorsed or recommended by ADB in preference to others of a similar nature that are not mentioned.

By making any designation of or reference to a particular territory or geographic area, or by using the term "country" in this publication, ADB does not intend to make any judgments as to the legal or other status of any territory or area.

Please contact pubsmarketing@adb.org if you have questions or comments with respect to content, or if you wish to obtain copyright permission for your intended use that does not fall within these terms, or for permission to use the ADB logo.

Corrigenda to ADB publications may be found at http://www.adb.org/publications/corrigenda.

Notes:
In this publication, "$" refers to United States dollars.
ADB recognizes "China" as the People's Republic of China, and "Korea" and "South Korea" as the Republic of Korea.

Cover design by Rocilyn Lacay.

# CONTENTS

# TABLES, FIGURES, AND BOXES

## BOXES

# FOREWORD

With the relentless innovations and exponential expansion of digital technologies, we are witnessing an awe-inspiring transformation of economies and societies. Digital services, products, and solutions are making us more efficient, enabling us to engage across barriers of language and space, expanding our access to knowledge, generating insights through data analytics, and augmenting our human capabilities. No wonder digitalization is reshaping social interactions, political processes, business practices, business models, markets, workplaces, leisure, and even visions of the future.

It is natural that these digital disruptions are creating challenges as well. At the Asian Development Bank (ADB), we are familiar with both the benefits and risks of development interventions, including their unintended consequences. In our operations, we carefully consider different scenarios and apply safeguards policies and risk management procedures to ensure that we are making optimal choices relative to all possible options and outcomes.

When it comes to digital solutions, the choices are less obvious, and the associated digital risks tend to be less widely known. Emerging technologies are particularly challenging to assess for potential risks. Novel solutions are often marketed with great emphasis on benefits, while the known and unknown negatives are downplayed, making it difficult to tell hype from reality.

With artificial intelligence, the risk landscape in digital domains has become complex, ever changing and therefore hard to grasp. And yet, to understand the context of ADB's developing member countries, and leverage technological advancements for our projects and ensure they achieve the intended impacts, ADB and its counterparts need to fully appreciate, if not apprehend, the risks. This primer is meant to raise awareness and provide readers with both a knowledge foundation and assessment tools that can be used to better assess digital risks. Cognizant of the speed of digital change, ADB plans to periodically add new topics and update content in future editions.

In summary, as ADB expands its investments in digital solutions within its portfolio, it has also increased its internal capacity to leverage new opportunities for digital development. We are fully committed to enhance our development impact while managing associated risks—including climate, environment, and disaster risks. I hope that through this primer the reader will more fully appreciate the complexity of digital risks relative to remarkable digital opportunities, and thus contribute more effectively to equitable prosperity and sustainable development in Asia and the Pacific.

**Bruno Carrasco**
Director General, Climate Change and Sustainable Development Department
Asian Development Bank

# ACKNOWLEDGMENTS

The preparation of this primer was financed by the Asian Development Bank (ADB) regional technical assistance project Digital Development Facility for Asia and the Pacific, which is cofinanced by the Republic of Korea e-Asia and Knowledge Partnership Fund.

ADB would like to acknowledge lead author Klaus Tilmes, senior consultant, ADB, who assembled a vast research base in the process of his research, and co-author Arndt Husar, senior public management specialist (Digital Transformation), Digital Technology for Development Division (CCDT), Climate Change Department and Sustainable Development Department, ADB, who also facilitated the linkages with work of the Interdepartmental Working Group on Assessing the Risk of Digital Technologies in ADB operations. This publication was developed under the overall guidance of Thomas Abell, director, CCDT, ADB. A diagnostic report prepared by TRPC Pte Ltd provided an initial framing for developing approaches to risk assessment of digital technologies in ADB operations.

The peer reviewers of this report were Eleonore Pauwels, director of the AI Lab at the Wilson Center and senior fellow at the Global Center for Cooperative Security; Prasanna Lal Das, digital economy and policy expert; and Assaad Sakha, senior risk management specialist, Office for Risk Management, ADB. Masatake Yamamichi, digital technology specialist (cybersecurity and data privacy) and Byeongjo Kong, digital technology specialist (data analytics and big data) provided additional comments.

Jess Macasaet copyedited, Lawrence Casiraya proofread the report, Rocilyn Lacay created the cover art and inside page graphics, and Joe Mark Ganaban typeset the final publication. Carmela Fernando-Villamar, digital technology officer, CCDT and Genny Mabunga, senior operations assistant, CCDT, provided administrative support.

# ABBREVIATIONS

| | |
|---|---|
| ADB | Asian Development Bank |
| AI | artificial intelligence |
| CCPA | California Consumer Privacy Act |
| CMM | Cybersecurity Capacity Maturity Model for Nations |
| COVID-19 | coronavirus disease |
| DE4A | Digital Economy for Africa/Digital Economy for All |
| DECA | Digital Ecosystem Country Assessment |
| DGRA | Digital Government Readiness Assessment |
| DMC | developing member country |
| DPA | data protection agency |
| DX | digital transformation |
| eAI | ethical AI |
| ESG | environmental, social, and governance |
| EU | European Union |
| GDPR | General Data Protection Regulation (EU) |
| ICT | information and communication technology |
| IEC | International Electrotechnical Commission |
| IoT | Internet of Things |
| IRM | integrated risk management |
| ISO | International Organization for Standardization |
| IT | information technology |
| ITU | International Telecommunications Union |
| NIST | National Institute of Standards and Technology |
| OECD | Organisation for Economic Co-operation and Development |
| OHCHR | Office of the United Nations High Commissioner for Human Rights |
| PII | personally identifiable information |
| PoW | proof-of-work |
| PRC | People's Republic of China |
| RMF | risk management framework |
| SMEs | small and medium-sized enterprises |
| USAID | United States Agency for International Development |

# EXECUTIVE SUMMARY

Digital technologies have emerged as the most transformative force of our time. As these technologies have become widely accessible, the unprecedented speed of their diffusion, innovation, and disruption has promised tangible benefits—ranging from access, inclusion, and economic opportunities to transparency, political participation, and resilience. The disruptions caused by the coronavirus disease (COVID-19) demonstrated that governments and businesses that had laid a digital foundation, with cloud-based data systems, modern security protocols, an agile workplace culture, and digitally enabled processes, performed far better than those that had not done so. In a post-COVID-19 world, these dynamics continue to play out and pose a growing challenge for developing countries as digital transformation accelerates globally.

Global spending on digital transformation is projected to reach $3.4 trillion by 2026, with the Asia and Pacific region set to capture the largest market share. For most enterprises, digital tech initiatives center on (i) operational improvements in core infrastructure and business functions; (ii) innovating and scaling up of digital operations for research, manufacturing, and supply chain optimization; and (iii) reshaping customer experiences. For governments, the focus is on digitally integrating internal systems and operations, transitioning to a new generation of e-services for citizens and businesses, and building long-term resilience to anticipate and adapt to future shocks. Additional investments in essential digital infrastructure may include digital systems, payment platforms, affordable connectivity solutions for underserved communities, and data and computing environments for education and health facilities. The potential opportunities are transformative.

Yet, the promise of digital transformation (DX) can often be accompanied by risks affecting public and private organizations of any type, size, and maturity, along with individuals and communities. Digital risks can be defined as the risks associated with the creation, delivery, and use of digital technologies, processes, and services that are deployed to achieve operational efficiencies, scale new business models, or deliver new services to customers or the public. The combination of technologies, data, business processes, and strategies can have undesirable and, at times, unforeseen consequences. Digital risks, such as cyber, third-party, business continuity, data protection, artificial intelligence (AI) ethics, and human rights risks, create uncertainties, may cause financial and reputational losses, and can derail DX initiatives. For public sector agencies, digital risks may involve the inability to deliver services and the accompanying erosion of social cohesion and trust in institutions, a decline in competitiveness, and/or a loss in revenues.

Assessing the opportunities and risks associated with digital technologies is a crucial first step to successful digital development programs. Rather than treating DX as a sprint with a fixed end point, it needs to be understood as an evolutionary process geared toward continuous learning and adaptation. These programs should establish a fact base and engage with government agencies, the private sector, and civil society throughout the development cycle to help inform strategic decisions, generate co-designed solutions, and build leadership and implementation capacity. Many adopt widely used diagnostic frameworks that combine horizontal, whole-of-government, and whole-of-society perspectives with mechanisms for strengthening collaboration and interoperability as part of the broader digital ecosystem. The digital risk assessment pilot of the Asian Development Bank (ADB) brings into focus a country's digital readiness as well as the broader policy, regulatory, institutional, and innovation landscape.

Currently, estimates indicate that more than 70% of DX initiatives are falling short of their objectives. Organizations are struggling with a lack of alignment with expected business outcomes, missing proof of concept, absence of change management strategies to build internal support, and failure to anticipate or rapidly respond to changes in regulations and/or user expectations. Meanwhile, public sector efforts to advance digital initiatives are lagging behind private sector programs, with progress being held back by legacy technologies, lack of digital skills, slow user adoption, and complex success metrics.

As DX initiatives are becoming increasingly important to an organization's life cycle, choosing the wrong hardware, software, and platform solutions can be costly. The choice of technologies should follow a strategic plan and step-by-step process focusing on the *why*, the *what*, the *what for*, the *how*, and the *when*; *trade-offs*; and the *what next*. Most organizations can improve the results of DX initiatives and manage unexpected digital risks with the help of a well-thought-out vendor selection and engagement strategy. Depending on project size and complexity, criteria other than price merit close attention, such as the nature of vendor involvement in the solution design, post-implementation services, and prior experience with digital risk management.

Equally important, since many development programs have implementation horizons of 7–10 years, they are susceptible to technological changes. Technology solutions that are designed "into" projects need to be informed by an over-the-horizon view as advancements can quickly overtake the initial project design, saddling countries and enterprises with obsolete technologies and unsustainable debt. Future thinking and technology foresight can be important tools for development practitioners to address this fundamental challenge. Multilateral development banks (MDBs) and development partners can play a crucial role in supporting an enabling environment through project preparation and capacity building, and also helping to mobilize private capital through blended finance, investment funds, pooled procurement, and guarantee instruments.

As organizations accelerate investments in new digital products and services, decision-makers are demanding greater visibility into the connections between strategy, operations, and technology. Most organizations already have risk oversight functions in areas like cybersecurity, information security, and regulatory compliance. But as digital risks grow in complexity and interconnectedness, traditional, internally focused, compliance-driven, and siloed risk management approaches are failing to keep pace with changing business strategy because they lack focus on emerging and atypical risks and are unable to embed risk mitigation and compliance activities across the organization. For example, what once may have been a sufficient mitigation approach in one part of the organization, such as information technology (IT) security, may now fall short of providing the required level of risk monitoring, prevention, and response in other areas, such as environmental, social, and governance (ESG) performance; supply chain continuity; or third-party risk. Going forward, the challenge for governments, business leaders, and risk practitioners will be to shift risk management philosophy toward a more proactive and integrated approach.

Effective **digital risk management** involves anticipating all possible digital risks so that preventative steps can be taken to safeguard ongoing operations or the onboarding of a new technology. Bridging these gaps between strategic, operational, and technological risk domains will require an integrated risk management (IRM) approach that focuses on the following:

- **Performance.** How does the leadership manage and measure the organization's overall strategic and operational performance in key areas such as digital transformation, third-party networks, and sustainability?
- **Compliance.** Are risk teams set up to comply with the expected increase in new laws and regulatory mandates? Are they able to monitor and remediate noncompliance events (e.g., data protection, labor relations, ESG) without delays?
- **Risk assurance.** How does the organization know it is mitigating priority risks in the right way? Has the organization (re)set its digital risk appetite and established meaningful risk metrics that are monitored continuously?
- **Resilience.** Is the organization able to quickly identify, respond to, and recover from a digital risk event (e.g., supply chain failure, cyber attack, system outage)? Are IRM teams aware of what is most critical to preserve and restore business continuity (e.g., through scenario exercises, digital inventories)?

Risk functions alone cannot single-handedly manage risk. It is critical to integrate them with other assurance functions by sharing data and insights, as well as engaging with operational teams. A unified risk framework helps to define the roles of each assurance group (e.g., internal audit, risk management, compliance, etc.); clarify their working relationships; and combine workflows, threat intelligence, risk metrics, and technology platforms into a single organization-wide risk assessment.

Digital risk management differs from traditional business risk management in several respects. A first distinction concerns the rapid, location-independent diffusion of risk. A ransomware attack can shut down an entire municipality for days or even weeks. A second characteristic involves the growing interconnectedness of internal and external business entities. A restaurant chain, for instance, can lose vast amounts of business if its third-party delivery partner changes its business model. A third difference that sets digital risk management apart is the expansive role of digital assets and data. Unlike traditional businesses with limited assets, employees, and partners, digital business models need to manage a very large number of assets, such as Internet of Things (IoT) devices or autonomous systems driven by algorithms that can make decisions and learn behaviors that organizations cannot predict. Finally, digital risk management needs to contend with the fact that digital business models have an enhanced ability to reinforce socioeconomic patterns and biases on a massive scale.

Managing the entire spectrum of digital risks at once is neither efficient nor effective. Not all risks have the same level of impact on desired outcomes. Instead, organizations should focus on the most critical risk areas threatening their health. For many organizations, cyber attacks, third-party risks, data leaks, and ethical flaws in the design and deployment of new products and services have the most negative repercussions and spillovers into other digital risk areas. Subject to country context, this assessment will need to be incorporated into the design of development programs as well.

**Cyber risks.** Cybercrimes top the global risk barometer. Converging technology platforms and interfaces connected via an increasingly decentralized and fragmented internet are creating a more complex cyber threat landscape and a proliferation of failure points. A web of profit is connecting the $6 billion underground cybercrime economies across the world and is impacting nation-state conflicts in online environments. The COVID-19 pandemic has exacerbated these threats through an expansion of ransomware attacks, supply chain disruptions, and financial fraud. As attackers continue to perfect the use of AI and other technologies, the end-to-end attack cycle will shrink from weeks to days to hours.

The proliferation of cyber attacks is outpacing societies' ability to effectively prevent and manage them. No organization is immune: federal, state, and municipal governments; schools and hospitals; and businesses, ranging from small enterprises to global corporations, have become targets. Companies and government agencies are frequently at a loss to identify and manage digital risks—yet are facing stiffer compliance requirements due to growing privacy concerns, high-profile breaches, and complex cross-border data flow regulations. While the short-term focus is on containing the impact of data breaches and intellectual property theft, organizations are concerned with the long-term damage to trust and reputation.

Cybersecurity risk assessments are increasingly replacing checkbox compliance routines to serve as the foundation of a risk management strategy. Cybersecurity maturity models can provide road maps for countries and organizations to measure, assess, and enhance cybersecurity by drawing on best practices and adapting a continuous improvement approach. All maturity models are structured along five pillars: physical security, people security, data security, infrastructure security, and crisis management.

Effective cybersecurity requires adequate policies and strategies, along with institutions that have the human and material resources to follow through on implementation. Governments, civil society, media outlets, and the private sector are well advised to incorporate cybersecurity into all operational aspects, including enterprise architecture, procurement, supply chains, and contracting agreements. Looking ahead, the next generation of cyber risk capabilities will require organizations to adopt new defensive cybersecurity capabilities as remote work and on-demand data access grow more common (e.g., zero trust architecture, behavioral analytics); address adversarial AI attacks that can reverse engineer AI models or inject manipulated data; and incorporate security-by-design approaches.

Until a few years ago, the niche nature of cybersecurity confined these issues to a small group of experts and stalled the integration of cyber capacity building into development agendas. This is changing. An important first step to bridge this gap between development and cybersecurity communities is to re-frame the narrative for cybersecurity capacity building in terms of digital resilience, safety, trust, sustainability, and risk management, and to link security with sustainable economic development and digital human rights.

Development organizations are beginning to embed defensive cyber capabilities into programs and projects. Securing critical infrastructure sectors—such as energy, IT and communications, transportation, health, defense, finance, commerce, emergency services, water and dams, and chemical installations— is a prerequisite for a functioning economy. Recent initiatives call for integrating cybersecurity and cyber capacity into digital development programs to achieve better outcomes and recognize that failure to address cyber risks may lower public trust in digital technologies. For development organizations, the message is clear: cyber incidents affecting development programs can cause long-term damage to their reputation and effectiveness in the partner country and around the globe.

**Third-party digital risk management.** Third-party digital risks are on the rise. As organizations increasingly rely on third-party contractors and supplier networks, they are exposed to a variety of legal, reputational, regulatory, and financial risks. Recent mega breaches have been caused by third parties having been granted too much privileged access to sensitive information; data exposures due to misconfigured cloud installations; or methodical "hack one, breach many" approaches that start by attacking a third-party vendor (e.g., SolarWinds) serving multiple organizations. Another risk dimension is the renewed regulatory focus on ESG issues, which is forcing organizations to consider not only their own footprint, but also to conduct appropriate risk assessments of their third parties' and suppliers' social impact. For instance, recently adopted supply chain laws by the European Union (EU) and Germany require companies to conduct environmental and human rights due diligence along their entire supply chain or face substantial fines.

How can third-party digital risk management (TPRM) programs comply not only with the flurry of new regulations but also cope better with emerging risks and strengthen their supply chains? With many contingency plans having failed to deal with the growing threat landscape, organizations are starting to see TPRM as an essential part of business resilience. TPRM programs need to move beyond their departmental silos such as procurement, compliance, IT security, risk, and data privacy, to a cross-functional approach of monitoring and managing third-party relationships. Fulfilling these requirements calls for a comprehensive solution and significant operational changes. Programs are now expected to monitor multiple risk domains including cyber security, data privacy, anti-bribery and corruption, and ESG issues. Programs also need to extend deeper into supply chains to address these risks, paying closer attention to fourth parties and beyond.

An effective third-party risk management framework and process that can be integrated with an organization's overall risk management requires (i) strategic analysis; (ii) engagement with third-party vendors and their suppliers; (iii) remediation aimed at evaluating and vetting third parties' security practices based on an acceptable level of risk; (iv) following remediation, a decision about whether to onboard the vendor or choose alternative suppliers; and (v) monitoring and reporting. Organizations can rely on several widely used risk management frameworks (National Institute of Standards and Technology [NIST], International Organization for Standardization [ISO]) to benchmark their TPRM programs.

**Privacy risk and data protection.** Countries cannot fully reap the benefits that digital tools and data integration offer without creating and maintaining public trust. To build and maintain trust, governments are playing a key role, be it by setting clear and enforceable rules to protect citizens' rights, embedding accountability and transparency in public data systems, or taking steps to curtail the market power of big tech companies. The challenge facing governments is how to establish rules that protect citizens from harm while at the same time encouraging innovation.

Privacy legislation continues to be very well received around the world. Globally, most countries have enacted privacy laws, driven by the catalytic effect of the EU General Data Protection Regulation (GDPR); growing awareness of the risks of data misuse; the desire to create an enabling framework for responsible innovation; and, more recently, the need to meet donor standards on data protection. However, many low-or middle-income countries struggle to translate new data privacy laws into action due to the (i) complexity of existing data privacy frameworks, (ii) lack of legal skills and knowledge needed to comply with data privacy laws across jurisdictions, and (iii) shortage of funding for data protection agencies. Development agencies are well positioned to provide institutional support. For companies, data privacy has become a top-level business imperative since most consumers would not buy from an organization that does not properly protect its data.

Global debates about data governance are still nascent, have focused largely on issues of cross-border data transfers, and reflect the priorities of wealthier countries. As a result, the data policy landscape remains fragmented. Over the coming years, privacy and data protection programs will have to adjust to a proliferation of new regulatory requirements. Prompted by these constant regulatory changes, organizations will need to evolve their privacy programs continuously to meet new requirements.

A privacy impact assessment (PIA) is useful for identifying and evaluating specific privacy risks. It can help organizations weigh the benefits of processing data against the risks; ensure compliance with applicable legal, regulatory, and policy requirements for privacy; and determine an appropriate response. Regulatory authorities have published detailed guides, process maps, and use cases for undertaking privacy impact assessments. At the same time, new data collaboratives between MDBs, foundations, and private technology companies have developed scalable solutions to global problems, including heat maps, traffic safety, and disaster response.

**Ethical AI risks.** AI applications and large language models, which are playing a significant role in digital transformation, can create new content in response to user requests in the form of text, code, voice, images, and videos. Amidst the hype triggered by the rapid adoption of these tools worldwide, leaders in tech, industry, academia, and civil society are issuing urgent warnings about the risks of AI and are calling for regulatory guardrails imposed by governments. This requires a holistic ethical framing to avoid unwanted developments such as undermining privacy, creating unfair outcomes, losing human control over increasingly intelligent systems, and preventing malicious and/or dual-use applications of generative AI.

Recent developments in AI regulations in the People's Republic of China (PRC), Europe, and the United States are instructive and hold important lessons for policymakers and development agencies.

The PRC has taken an early lead in adopting regulations governing companies' use of consumer-facing algorithms in online recommendation systems, requiring that such services be moral, ethical, accountable, transparent, and "disseminate positive energy." The regulation mandates companies to notify users when an AI algorithm is shaping which information they see and prohibits algorithms that use personal data to offer different prices to consumers. The PRC's recent proposal for regulating generative AI strongly favors protecting society (and the government) from risks posed by AI systems and tech companies, squarely shifting regulatory powers to the government.

The European Commission is poised to adopt a far-reaching regulatory framework to protect the "right to non-discrimination, freedom of expression, human dignity, personal data protection, and privacy." The riskiest AI systems would be banned outright as they are deemed to present "unacceptable" risks to public safety and human rights by (i) deploying cognitive behavioral manipulation of people or specific vulnerable groups, (ii) enabling "social scoring" by classifying people, and (iii) facilitating real-time and remote biometric identification. The most significant impact of these new rules is that they will apply not only to providers of AI systems based in the EU, but also to businesses outside the EU that provide AI-based products and services used inside it—effectively setting a global standard.

Meanwhile, the AI guidelines under discussion in the United States are voluntary, non-prescriptive, and focused on changing the culture of tech companies. For instance, The White House called for public assessments of generative AI systems to prevent the new technology's misuse. Following extensive global consultations, the NIST released the AI Risk Management Framework 1.0 (RMF), which assesses trustworthiness throughout the design, development, use, and evaluation of AI products, services, and systems. The reasoning is that trustworthy AI systems will help preserve civil liberties and rights and enhance safety, while creating opportunities for innovation.

What lessons can governments genuinely wanting to promote democratic values draw from these three approaches? First, many of the consensus AI principles often cited as supporting democratic values can just as easily be used for authoritarian purposes. Demonstrating that AI applications live up to principles such as "fair," "accountable," and "human-centered" will require resources and inclusive processes to collect input from diverse publics about what they do and do not want from new technologies. Second, many of the most consequential uses of AI systems, from the provision of critical services to policing and surveillance, are situated within government agencies. As a result, regulations that apply only to the private sector ignore some of the strongest potential negative impacts on democracy. Third, an embrace of AI systems that erode individual privacy, perpetuate discrimination, and widen inequality are likely to undermine public trust in government and businesses.

Many organizations are facing practical challenges from a risk management perspective to navigate these issues and guide the development of ethical AI (eAI) systems. Minimum requirements for an effective risk management program that anticipates and mitigates eAI risks include (i) the need to establish accountability for the decisions of one's AI system as it interacts with others; (ii) the adoption of governance principles, policies, and/or protocols, as well as continuous monitoring of their proper implementation; (iii) the ability to explain algorithmic decisions to end users and other stakeholders in nontechnical terms; and (iv) providing transparency about how and why an AI-system made a specific decision, and, if such a system caused harm, how to discover the root cause.

Technology practitioners, business groups, and policymakers rely on tool kits to help guide and support their AI ethics work. AI ethics tool kits can play an important role in promoting more ethical and fair approaches by identifying how AI systems can have harmful effects or by anticipating potentially harmful impacts on the organization itself that is developing or deploying the AI systems. To strengthen the potential impact of these tool kits, organizations should identify upfront the stakeholders and use cases in which these AI systems will be deployed and pitch AI ethics as a problem for collective action among multiple groups of stakeholders, rather than as an agenda for individual expert-practitioners.

**Digital risks for human rights and vulnerable groups.** Digital technologies offer new opportunities to ensure the protection of digital human rights. All too often, however, digital tools are also used to infringe on human rights, be it through data privacy breaches, mishandling of digital identities, mis- and disinformation, or online harassment, especially against women and girls. For refugees and migrants in need of humanitarian assistance, the use of digital IDs and biometrics has raised repeated human rights concerns around surveillance practices, informed consent, and choice. MDBs would be well advised to incorporate human rights risk factors associated with the data cycle (collection, storage, use, and re-use) into their risk assessments to ensure the protection of vulnerable groups.

Over the last decade, internet freedoms have declined globally. Drawing on 70 country assessments by Freedom House, global norms have shifted toward increased state intervention. Many governments have enlisted the private sector for contract tracing, control of online content, and targeted surveillance. Another benchmarking study gave major platform companies negative scores for their failure to disclose their human rights due diligence policies and practices and to protect users' privacy and freedom of expression. Recent decisions by several major tech companies to lay off in-house ethics and content moderation teams while at the same time accelerating the push into AI products are also worrying.

In a hopeful sign, some two dozen countries saw improvements in internet freedoms, with civil society organizations playing a crucial role in improving legislation, developing media resilience, and ensuring accountability among tech companies. Growing advocacy in support of digital human rights is also coming from institutional investor groups that seek to comply with ESG standards. Cities have recently kicked off initiatives that uphold a human rights-based approach for the digitalization of public services and their inclusive, equitable, and data-protected provision.

Overall, the pandemic has deepened the digital divide for vulnerable groups, who risk falling further behind unless steps are taken to ensure that the benefits of digitalization are fairly distributed across the whole of society. In the case of school-aged children, for instance, most EdTech products were offered at no direct financial cost during the COVID-19 pandemic. While this allowed governments to offload the actual costs of online education, students were unknowingly forced to pay for their learning with their freedom of expression as their personal data was passed on to ad-technology firms to influence buying preferences and lifestyle choices. Students' exposure to harmful digital content, increased social media use, and cyberbullying can have negative consequences for their personal development, reading skills, and confidence levels, especially for girls.

Clear international guidelines and compliance frameworks are needed to protect the rights of children in the digital environment. Educational AI systems, which claim to offer personalized pathways for learning, should not be deployed without an ethical framework that explicitly considers issues such as fairness, transparency, accountability, bias, and agency. Recommendations against the profiling of children on digital platforms and the adoption of child rights impact assessments are gaining traction globally.

**Sustainability risks.** The expansion of digital systems is an integral part of the broader sustainability discussion. Digital technologies can enable decarbonization across sectors and promote circular and shared economies, resource and energy efficiency, the monitoring and conservation of ecosystems, the protection of the global commons, and sustainable behaviors.

However, this is not an automatic process. The impact of digital technologies on global sustainability is widespread—as are their origin points. The supply chain life cycle for digital products begins with the extraction of metals and rare earth minerals, which tend to be concentrated in countries beset with political instability, weak governance, and severe environmental damage. On the downstream side, the lifetime of digital hardware is driven by business models that promote frequent equipment upgrades, making e-waste the world's fastest growing waste stream, of which less than 20% is recycled.

Given growing demand for digital devices and services, what policy actions are being considered? There are growing pressures to improve the governance of information and communication technology (ICT) supply chains, especially regarding the extraction of rare earth minerals and metals, the recycling of e-waste, and the safe disposal of toxic material. Evolutionary design and circular economy models could extend the life span of devices and enable the replaceability of key components.

Another area of concern is the energy consumption of digital services. Data centers are the information backbone of an increasingly digitalized world. Demand for their services is rising rapidly and accounts for around 1% of global electricity consumption. Will the next doubling of data center compute instances continue past trends of decreasing energy intensity? Possible policy actions include the promotion of energy efficiency standards, a shift to cloud services, and increased investments in renewable energy. Similarly, the mining of crypto currency requires energy- and computer-intensive operations. Bitcoin's annual electricity consumption is comparable to that of the Philippines or Pakistan. Crypto mining has become an issue for crypto hubs in Europe, Asia, and the US, where authorities are concerned that crypto's energy use is undermining their efforts to move away from fossil energy sources.

**Digital resilience.** The COVID-19 pandemic has prompted a global reassessment of digital resilience and the increasing reliance on digital systems to cope, recover, and adapt to a world characterized by increasingly frequent, unpredictable, and unprecedented disruptions.

A fundamental mindset shift away from a narrow view of risk management focused on a small set of well-defined events toward a holistic, people-centered approach to digital resilience is needed. This broader perspective requires engagement with the entire digital ecosystem—the stakeholders, systems, and enabling environments that together make it possible for people and communities to be able to use digital technologies to pursue economic and social opportunities in a safe and secure manner. Currently, the digital ecosystem is fragmented with a lack of cross-sector engagement and collaboration to identify and address the risks of digital technologies and establish mechanisms for joint actions that respect inclusiveness and digital human rights. The lack of global standards for interoperability and the absence of legal and ethical accountability systems to protect people's well-being against malicious use of data and technologies are obstacles to fully leverage these opportunities.

Despite the increasing urgency for strengthening digital resilience, most public and private sector organizations are struggling to define their resilience agenda and approach, hampered by the additional cost of building redundancy and a lack of a universal metric to assess the effectiveness of investments. Instead of restricting the resilience agenda just to the physical and technical aspects, digital resilience should extend across sectors and activities increasingly dependent on digital technologies, and therefore, vulnerable to digital disruptions that could impact communities, organizations, and entire countries and regions. Digital technologies can help strengthen the resilience of these areas through enhanced data collection, processing, diffusion, and delivery capabilities. An inspiring example is the mRNA Vaccine Technology Transfer Hub: a coalition of 15 middle-income countries, including Argentina, Bangladesh, Brazil, India, Indonesia, Senegal, South Africa, Ukraine and Viet Nam, has come together to develop and produce mRNA vaccines. Beyond addressing vaccine inequities, this initiative is catalyzing job creation and economic growth.

In this connected world, it is no longer sufficient to attend only to what is within a specific jurisdiction, sector, or expertise. Governments need to find new ways to collaborate across silos and borders, and adopt favorable regulations and incentives for embedding resilience in all future policymaking. Corporations able to move beyond a narrow focus on risk and controls to a longer-term strategic view will be able to turn resilience into a competitive advantage in times of disruption. Crucially, this is the moment for development agencies to embrace the digital resilience agenda for a sustainable and equitable future.

**Technologies empowering economies.** Embracing the reliance on digital technologies for efficiency and connectivity (photo by ADB).

# INTRODUCTION

The growing reliance on digital technologies is altering economies, societies, governance, and interpersonal relationships. Digital transformations offer opportunities to scale innovations, launch entirely new business models, and promise greater efficiencies and convenience for organizations and the public alike.

As countries adopt a greater number of digital systems and tools, however, new vulnerabilities continue to emerge. Digital risks represent one of the fastest growing, most pervasive risks for governments, enterprises, civil society organizations, and even entire countries. They encompass a diverse set of unwanted—and often unexpected—outcomes inherent in digital products, services, and processes that are associated with digital transformation. This new wave of technological risks—from digital inequality to cross-border cyber attacks and disinformation—are recognized as critical threats, even as current risk mitigation efforts fall short of meeting the challenge.[1]

This primer introduces the concept of digital risk as a crucial dimension affecting development assistance in the age of digital transformation. The report offers opportunities to integrate digital risk considerations throughout the programming cycle and highlights appropriate mitigation strategies. It is primarily intended for development practitioners to raise awareness and provide guidance on digital risks and ethical concerns as they relate to the use of digital technologies in ADB's operations—be it through lending operations, advisory services, or knowledge and learning activities.

To set the stage, the primer outlines key digital assessment frameworks in **Chapter 1** that have proven useful for diagnosing strengths, gaps, opportunities, and risks at country level and within government. These diagnostic tool kits help set strategic direction, facilitate engagements with governments and stakeholder groups, and provide a fact base for digital country risk assessments.

**Chapter 2** offers an overview of digital transformation initiatives and key implementation challenges. Given concerns about technology selection and obsolescence, the chapter outlines criteria and methods for technology selection and forecasting.

**Chapter 3** provides an organizing framework across the spectrum of digital risks and proposes an integrated approach for managing them. The chapter concludes with observations on the maturity and capabilities of digital risk management.

---

[1] World Economic Forum. 2022. Global Risks Report 2022. https://www.weforum.org/reports/global-risks-report-2022/.

Against the background of increasing cyber threats by state and non-state actors, **Chapter 4** outlines cybersecurity frameworks, maturity models, and prospective cyber security capabilities. Reflections on the growing relevance of cybersecurity for development assistance set the stage for recommendations for integrating cybersecurity into development assistance at strategic and operational levels.

The increasing reliance on third-party contractors and supplier networks has added a new dimension to the risk landscape and exposes organizations to a variety of legal, reputational, regulatory, and financial risks. **Chapter 5** outlines gaps and vulnerabilities in the selection, onboarding, and monitoring of third parties and offers suggestions for mitigating the resulting risk exposures.

Data flows can create economic benefits and societal value for individuals and communities, provided trust is built and sustained between the parties who share, collect, and use data. A convergence of changing privacy expectations on the part of citizens and consumers, digital market competition, and accelerating government action is ushering in new rules on data privacy and data sharing which are underpinning today's data economy. **Chapter 6** focuses on key frameworks, standards, and tools affecting privacy risk and data protection.

AI and large language models are playing a leading role in digital transformations and require a holistic ethical framing to avoid unwanted developments. Amid the hype accompanying recent advances in generative AI applications, leaders in tech, industry, academia, and civil society are issuing ever-more urgent warnings about the risks of AI and calling for regulatory guardrails. **Chapter 7** compares recent developments in AI regulations in the People's Republic of China (PRC), Europe, and the United States; outlines a field-tested approach for assessing and mitigating ethical AI risks; and provides an overview of AI tool kits and guidelines intended to tackle the rapidly evolving legal, ethical, and societal challenges associated with AI.

Digital technologies offer new opportunities to exercise human rights, but frequently are used to infringe on them, be it through data privacy breaches, surveillance, or online harassment. **Chapter 8** reviews policies and practices by governments and major platforms and highlights the exposure of vulnerable groups to digital risks.

The expansion of digital systems is an integral part of the broader sustainability discussion. **Chapter 9** introduces a supply chain life cycle for digital products and e-waste and discusses the energy consumption of data centers and cryptocurrency mining.

The COVID-19 pandemic has prompted a global reassessment of digital resilience and the increasing reliance on digital systems to cope, recover, and adapt to a world characterized by more frequent and severe disruptions. Learning from these crises, **Chapter 10** concludes with a set of proposals to shift from an ad hoc response to a holistic digital resilience strategy.

1 DIGITAL DIAGNOSTICS: METHODS AND APPROACHES

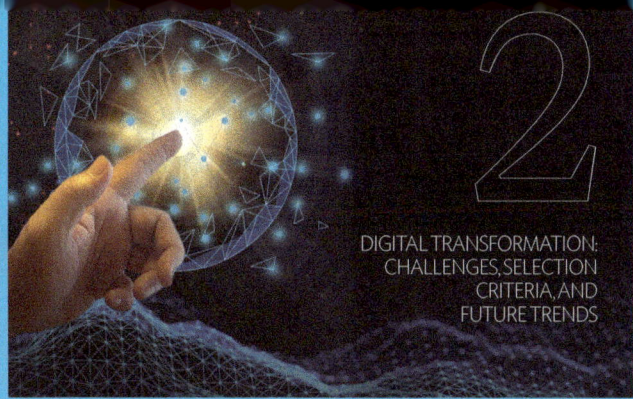

2 DIGITAL TRANSFORMATION: CHALLENGES, SELECTION CRITERIA, AND FUTURE TRENDS

3 DIGITAL RISK

4 CYBERSECURITY RISKS

5 THIRD-PARTY DIGITAL RISK MANAGEMENT

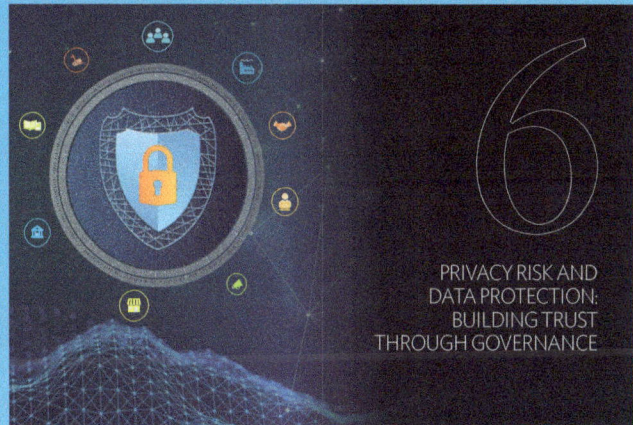

6 PRIVACY RISK AND DATA PROTECTION: BUILDING TRUST THROUGH GOVERNANCE

7 ETHICAL AI RISKS

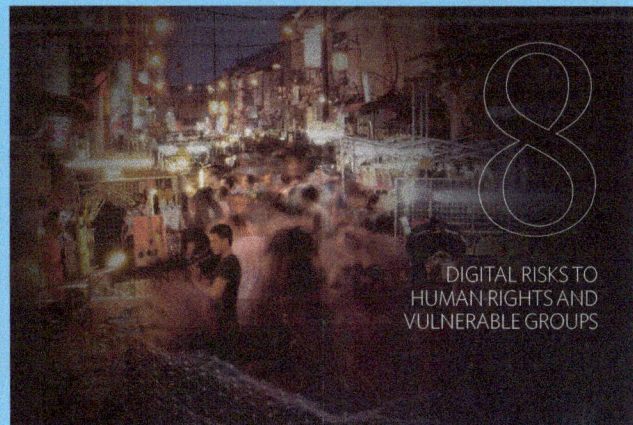

8 DIGITAL RISKS TO HUMAN RIGHTS AND VULNERABLE GROUPS

9 SUSTAINABILITY RISKS

10 DIGITAL RESILIENCE

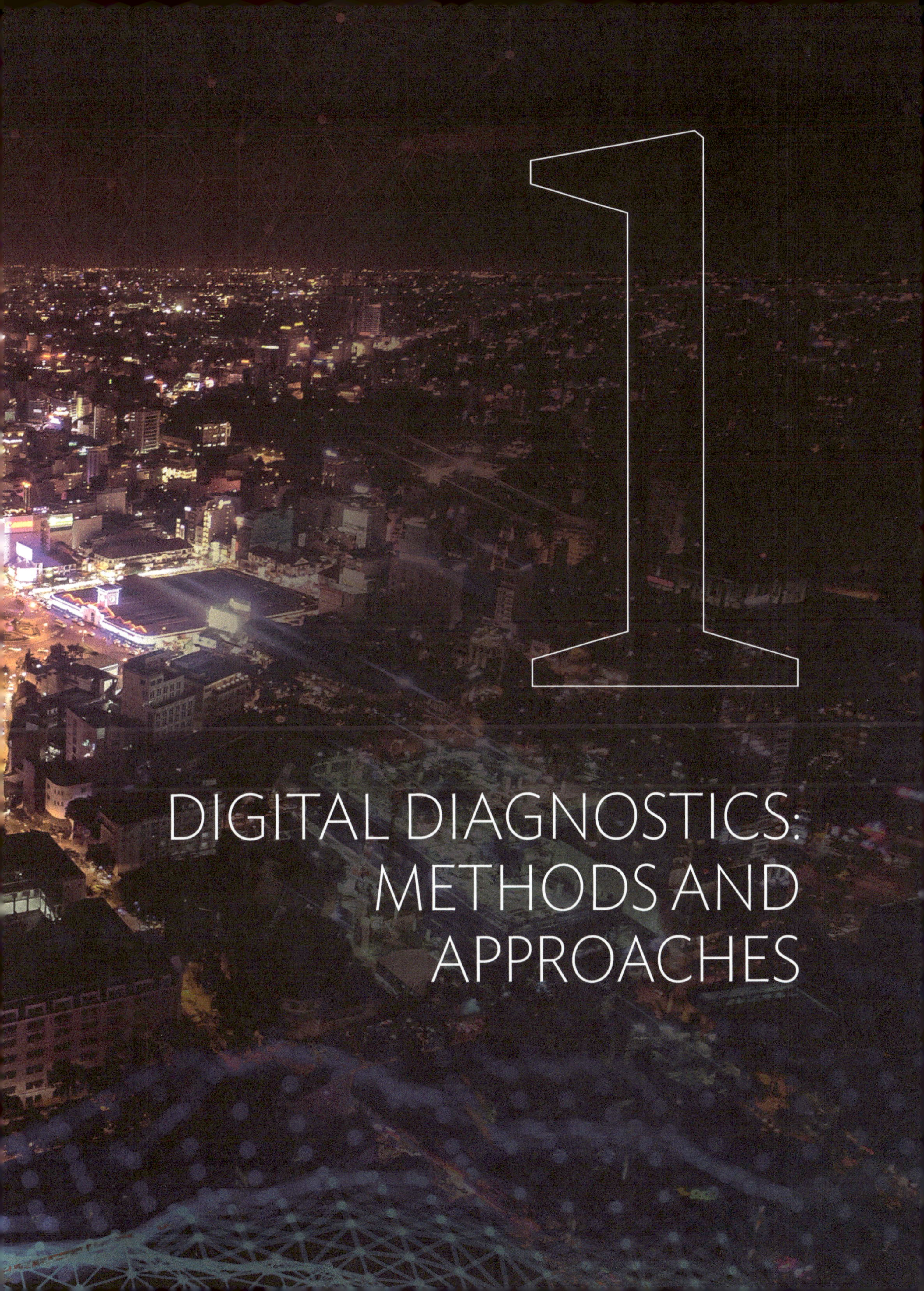

# 1

# DIGITAL DIAGNOSTICS: METHODS AND APPROACHES

Digital technologies have emerged as the most transformative force across societies and economies over the last 2 decades. As these technologies have become more accessible, their unprecedented speed of diffusion, innovation, and disruption have promised tangible benefits from digitalization, ranging from inclusion, equality, and economic opportunity to transparency, shared prosperity, and political participation. Yet, they also have also been accompanied by significant risks to data privacy, security, and trust through misinformation, surveillance, censorship, and cybercrime. It is therefore imperative to understand the potential opportunities and risks associated with digital technologies within a country's digital ecosystem for both public and private sector institutions, as well as between countries.[2]

The development community has become enthusiastic about the prospects of digital technologies and data for development. Growing awareness of digital dividends at country level, supportive leadership, and a series of influential reports have increased demand for stepped-up engagement.[3] Starting a few years ago, international development agencies and think tanks began to pilot a set of diagnostic tools for assessing the digital economy and for addressing complementary reforms and investments.

This chapter will explore several widely used digital diagnostic frameworks which provide an overall context for the assessment of digital country risks that offer distinct but highly complementary perspectives on the digital foundation of economies, societies, and governments. The underlying methodologies have been adapted to countries at different digital maturity levels and translated into open-source tool kits to encourage wider application. The growing body of reference material helps identify emerging practices, taps into a wealth of qualitative and quantitative cross-country information, and serves as a learning platform for policymakers, stakeholders, and development practitioners.

While this broad overview presents the field of digital diagnostics from different angles, none of the three frameworks addresses all digital risk factors in their entirety or offers an in-depth analysis for specific risks. They represent the closest to a compromise between a very specific and an all-encompassing—and potentially unwieldy—360-degree assessment of all possible risks.

Effective tools for understanding digital technology risk at the country level will play a key role for policy engagements and the design and implementation of technical assistance and investment programs. For that purpose, the Asian Development Bank (ADB) and its foray into digital risk assessment brings into focus a country's digital readiness as well as key aspects of the policy, regulatory, institutional, and innovation landscape affecting the digital risk profile.

---

[2]   United States Agency for International Development (USAID) and Digital Frontiers. Digital Ecosystem Framework. https://www.usaid.gov/digital-development/digital-ecosystem-framework.

[3]   Examples include World Development Report 2016, UN High-level Panel on Digital Cooperation, Pathways for Prosperity Commission, World Development 2021.

# 1.1. Digital Economy for Africa Framework

The World Bank launched the Digital Economy for Africa (DE4A) initiative in 2018 with the goal of "digitally enabling every individual, business and government in Africa by 2030."[4] This pan-African initiative aims to enable all countries to work together with African Union member states, development partners, and the private sector to set up relevant programs and policies to catalyze Africa's digital transformation. The DE4A initiative is based on five core principles (comprehensive, transformative, inclusive, homegrown, and collaborative) and comprises five building blocks: digital infrastructure, digital platforms, digital financial services, digital entrepreneurship, and digital know-how.[5]

The DE4A Diagnostic Framework (Version 1.0) operates at three levels: macro policy foundations (competition, trade, finance, governance, etc.); digital enablers (digital leadership, infrastructure, platforms, policies, skills, finance); and sector transformations (vertical ICT applications in key economic sectors). These components and their interactions are the key elements for co-producing the economic impact of digital transformation, often referred to as digital dividends.

Building on the lessons emerging from the first dozen country reviews, Version 2.0 of the DE4A diagnostic tool introduced several changes, including (i) a clearer articulation of digital economy pillars and their interactions; (ii) a distinction between public and private digital platforms; (iii) the addition of crosscutting areas such as data privacy, cybersecurity, and gender; (iv) the recognition of capacity constraints in key regulatory areas; and (v) the sequencing of reforms to reflect country context and digital maturity (see Figures 1 and 2). Building on this experience, a newly released report on the status of the rapidly accelerating digital economy across South Asia offers fresh insights about the need for a trust environment and opportunities for regional collaboration (for cross-border connectivity, data flows).[6]

Experience with digital economy assessments in Africa and South Asia highlights several common challenges and lessons:[7]

- Digital diagnostics were often conducted in isolation of countries' overall development strategies. Ideally, the formulation of digital economy and country economic strategies should proceed interactively, as digital technologies can open new opportunities for deep economic transformation.
- The diagnosis of digital economies erred more on the side of strengths and opportunities, and less on the side of the accompanying risks, tradeoffs, and downsides of digitization, as well as the countries' capacity for managing these risks. The gaps in safeguards stand out when

---

[4] The initial DE4A designation with its focus on Africa has since been broadened into Digital Economy for All.

[5] The DE4A initiative is accompanied by a suite of diagnostic tools, including the Digital Economy Country Assessment tool; the Digital Infrastructure Assessment tool; the Digital Government Readiness Assessment tool; the Financial Inclusion Guidance Note for the Financial Sector; the Digital Entrepreneurship Ecosystem diagnostic; and the Digital Identification (ID) diagnostic tool. Digital Economy for Africa (DE4A) Diagnostic Tool and Guidelines for Task Teams Version 1.0. https://thedocs.worldbank.org/en/doc/320381611092376395-0090022021/original/DE4ADiagnosticTool.pdf.

[6] World Bank. 2022. *South Asia's Digital Opportunity: Accelerating Growth, Transforming Lives*. Washington, DC. https://openknowledge.worldbank.org/handle/10986/37230.

[7] N. K. Hanna. 2020. Assessing the Digital Economy: Aims, Frameworks, Pilots, Results, and Lessons. *Journal of Innovation and Entrepreneurship*. 9(1). p.16. https://doi.org/10.1186/s13731-020-00129-1.

## Figure 1: Foundational Elements of a Digital Economy

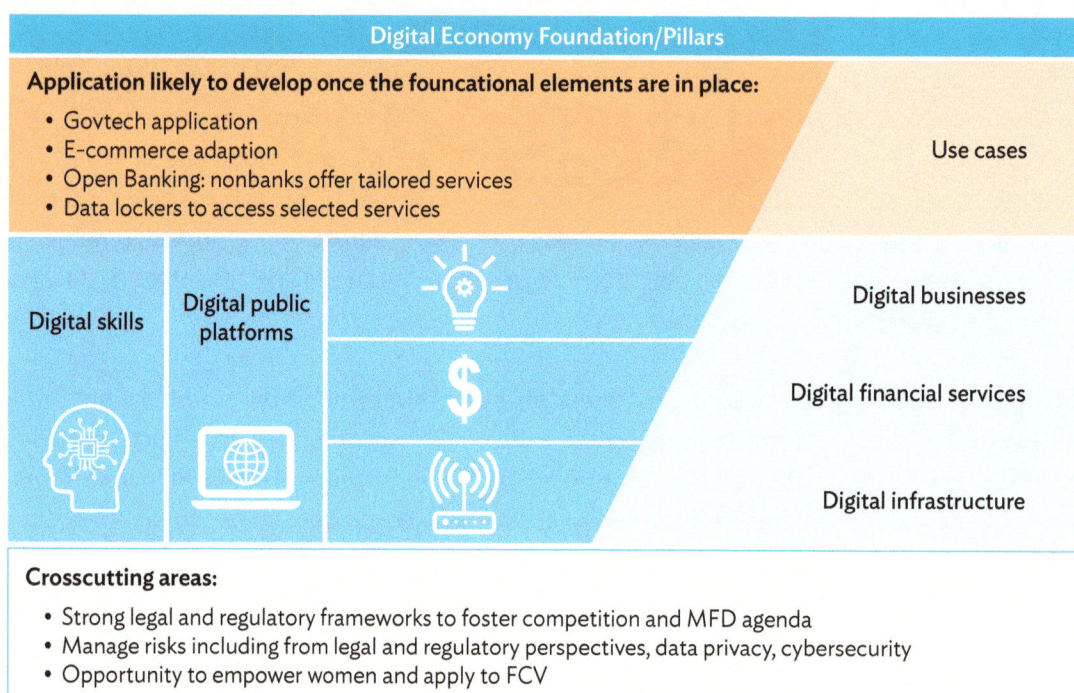

| Digital Economy Foundation/Pillars | | |
|---|---|---|
| **Application likely to develop once the foundational elements are in place:**<br>• Govtech application<br>• E-commerce adaption<br>• Open Banking: nonbanks offer tailored services<br>• Data lockers to access selected services | | Use cases |
| Digital skills | Digital public platforms | Digital businesses<br><br>Digital financial services<br><br>Digital infrastructure |
| **Crosscutting areas:**<br>• Strong legal and regulatory frameworks to foster competition and MFD agenda<br>• Manage risks including from legal and regulatory perspectives, data privacy, cybersecurity<br>• Opportunity to empower women and apply to FCV | | |

FCV = fragility, conflict, and violence; MFD = Maximizing Finance for Development.
Source: World Bank 2020: Digital Economy for Africa Country Diagnostic Tool and Guidelines for Task Teams.

## Figure 2: Assessment Areas Based on Countries' Digital Maturity

| | Nascent | Growing | Advanced |
|---|---|---|---|
| Digital infrastructure | Access to undersea internet cables, backbone networks | Backbone networks, data clouds, IXPs, privacy, and cybersecurity | 4G/5G networks, rural connectivity, Internet of Things |
| Digital platforms | Digital shared services, digital identity, and digital financial management | Digital government, open data, e-commerce | Mobile apps, AI applications, and software-enabled platforms |
| Digital financial services | Basic digital payments, e.g., person-to-person payments | Broad digital payments, e.g., business-to-person, government-to-person | Digital financial services, e.g., savings, credit insurance |
| Digital entrepreneurship | Talent development and business mentoring | Angel/seed financing, innovation centers, regional hubs | Venture financing, M&A, IPOs, BPO centers, local digital industry |
| Digital skills | Boot camps and digital skill trainings | Business/management skill training | Digital-savvy workforce |

Source: World Bank 2020: Digital Economy for Africa Country Diagnostic Tool and Guidelines for Task Teams.

examining the robustness of key protections in the processing of personal data.[8] Creating a rights-based approach to data protection that regulates data collection, processing, and use by third parties, and improving cybersecurity are critical steps for increasing trust in cyberspace. Most development agencies are actively recruiting scarce skills in these areas to respond to increased client demand for assessing and mitigating digital risks and adapting their fiduciary and safeguard responsibilities.

- Most diagnostics did not assess the implementation record of past strategies or the capabilities of existing institutions to implement proposed strategies. Digital transformation often calls for developing new digital leadership and institutions, mobilizing local digital capabilities, and strengthening digital governance.

- Few digital economy assessments sought to secure broad-based participation by including institutions representing small business, civil society, trade and professional associations, and vulnerable communities. The systematic tracking of distributional and empowerment impacts of new technologies can yield important insights on entrenched barriers to adoption and inclusion.

- In most pilot countries,[9] assessment indicators did not adequately capture adoption and effective use of digital technologies by public agencies, traditional industries, small businesses, and citizens. For most countries, there is significant scope to stimulate demand for innovative and locally tested digital solutions, especially those coming from local innovators and technology-oriented small and medium-sized enterprises (SMEs) that could be scaled up and integrated into a national digital economy strategy.

- The initial assessments did not provide adequate coverage of digital inclusion and income inequality and were too aggregated to capture digital-related income, gender, and geographic disparities—critical dimensions for digital economy strategies.

- Most digital economy assessments proceeded along vertical silos, mirroring the different digital economy components. Advancing economy-wide digital transformation requires a horizontal, whole-of-government approach. Economy-wide digital assessments are expected to improve collaboration between key sectors to deliver more integrated digital solutions and help break down ministry and sector silos.[10]

- The DE4A diagnostic program also demonstrated that modularity without a common framework and clear links between modules carries risks for both development agencies and client countries by losing sight of the whole and underestimating coordination requirements. This is particularly critical for country-level implementation, where the various components of the digital ecosystem must be aligned and coordinated. Integration of assessment tools is not an easy task. A modular digital tool kit should be customized to the needs of member countries and integrated into policymaking to facilitate client learning and capacity building.

---

8   World Bank. 2022. *South Asia's Digital Opportunity*. Washington, DC.
9   World Bank. DE4A Country Diagnostics Status. https://www.worldbank.org/en/programs/all-africa-digital-transformation/country-diagnostics.
10  International Telecommunications Union (ITU) Hub. 2021. G5 Regulation: The Digital Transformation Fast Lane. 17 November. https://www.itu.int/hub/2021/11/g5-regulation-the-digital-transformation-fast-lane/.

# 1.2. USAID's Digital Ecosystem Framework

The digital strategy of the United States Agency for International Development (USAID) goes beyond the narrow definition of a digital economy and seeks to achieve and sustain open, secure, and inclusive digital ecosystems. Its digital ecosystem framework brings together stakeholders, an enabling environment, systems, and culture, that, together, empower people and communities to use digital technologies to access services, engage with each other, and pursue economic opportunities. The Digital Ecosystem Country Assessment (DECA) serves to inform the development, design, and implementation of USAID's strategies, projects, and activities. Based on experiences gained through five pilot assessments in Colombia, Kenya, Nepal, Uzbekistan, and Serbia, the DECA is designed as a 5- to 6-month long research, interview, and dissemination process. A recently published DECA tool kit provides step-by-step guidance.[11]

USAID frames a country's digital ecosystem around three pillars: Pillar I on Digital Infrastructure and Adoption; Pillar II on Digital Society, Rights, and Governance; and Pillar III on Digital Economy; and four crosscutting themes of inclusion, cybersecurity, emerging technologies, and geopolitical positioning.

The distinguishing characteristic of USAID's framework is Pillar II, which focuses on how digital technology intersects with government, civil society, and the media and can shape the nature of digital risks. It is divided into three sub-pillars:

- Internet freedom refers to the ability of individuals to access the internet without obstacles, produce and consume content without censorship, and have their fundamental human rights respected online. Assessing internet freedom requires understanding which online rights are recognized, how these rights might be violated, and what institutions and processes govern the online space. *Digital rights*, for instance, include rights relating to data privacy and data ownership, protecting children from digital harms, and protecting women and girls from online gender-based violence. The digital rights topics also considers the extent to which private sector companies use human rights impact assessments to identify and prioritize human rights impacts. *Digital repression* can violate human rights through surveillance, censorship, social manipulation and disinformation, internet shutdowns, and online persecution of users. Digital repression is not limited to government actors; non-state and foreign actors can also deploy these techniques for political, social, and economic reasons. How these principles, norms, rules, and decision-making procedures are developed and applied, whether through a collaborative multi-stakeholder process or a state-driven approach, is captured under the notion of *internet governance.*

---

[11]   USAID. Digital Ecosystem Country Assessment (DECA) Toolkit: A How-To Guide for USAID Missions. https://www.usaid.gov/digital-development/deca-toolkit.

**Figure 3:** USAID's Digital Ecosystem Framework

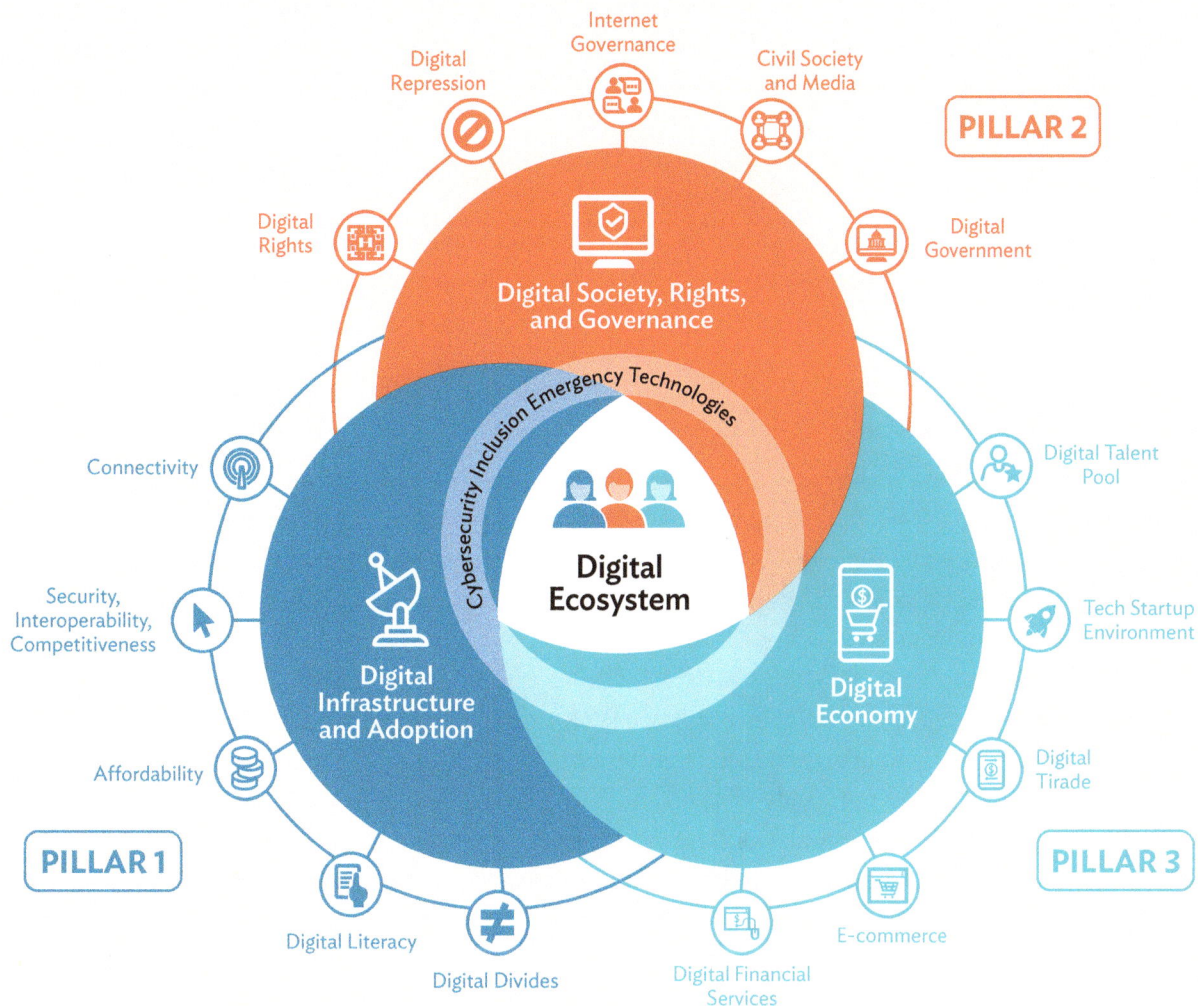

Source: USAID and Digital Frontiers. Digital Ecosystem Framework.

- Civil society and the media play an important role in a digital ecosystem by promoting the inclusion of diverse perspectives and calling attention to abuses. This aspect explores how these institutions serve a watchdog role in the face of declining online freedoms, how digital media is used and accessed, how social media is used for activism, and how the internet is used for political organizing.

- Digital government refers to the use of digital technologies to deliver government services (e.g., through digital ID systems, e-citizen portals); manage government systems (e.g., cloud storage, government data centers); and engage citizens and organizations (e.g., through feedback mechanisms, government innovation hubs). Effective digital government systems build public trust by respecting individual rights, especially by instituting guardrails to limit the use of technology by government and other actors for malign purposes. Governments can visibly constrain misuse by adopting and enforcing data governance policies and ethical guidelines (e.g., for the deployment of AI systems); using transparent procurement; and consulting impacted communities when deploying new technologies.

These issues of digital rights, society, and governance have cross;cutting implications for digital risks. For instance, national-level *cybersecurity* requires the protection of government data and IT systems, cross-border data flow agreements, and continued adaptation to new threats. Cyber attacks on government systems can decrease public trust in government processes and systems. Government cybersecurity institutions are critical but require adequate staff and funding. *Emerging technologies* can enhance targeted delivery of e-services but also pose potential risks in form of "deepfakes" and advanced surveillance systems. Foreign actors can use digital technologies to exert and broaden their *geopolitical influence* through commercial spyware or by undermining the stability of a government through disinformation. Mitigating digital risks must go hand in hand with the expansion of digital government.

## 1.3. Digital Government Readiness Assessment

A third diagnostic tool, the Digital Government Readiness Assessment (DGRA) Version 3.0,[12] helps governments assess their readiness to provide leadership and governance for digital transformation; apply digital technologies to improve administrative operations and services; invest in a shared digital infrastructure with strong cybersecurity measures; and provide laws and regulations for data privacy, consumer protection, digital identification, and cybersecurity.

Using both qualitative and quantitative analyses, the DGRA tool kit identifies strengths and weaknesses of current digital government efforts and proposes action steps to develop a comprehensive national digital strategy. The assessment considers the rising use of data by government and citizens; the increasing availability of Big Data; and the prevalence of networks, AI, and analytic tools for governments to become data driven. The tool kit has been used to diagnose country-specific digital government strategies and programs that can be financed by development partners.

The DGRA tool kit comprises nine foundational areas (Figure 4) for building open and agile digital infrastructure and operations: (i) Leadership and Governance; (ii) User-Centered Design; (iii) Public Administration and Change Management; (iv) Capabilities, Culture, and Skills; (v) Technology Infrastructure; (vi) Data Infrastructure, Strategies, and Governance; (vii) Cybersecurity, Privacy, and Resilience; (viii) Legislation and Regulation; and (ix) Digital Ecosystem. These foundational areas are surveyed through a total of 67 assessment questions. A subset of these questions, which directly touch on key dimensions of digital risks, is listed in Table 1.

---

[12]    World Bank. 2020. Digital Government Readiness Assessment Toolkit: Guidelines for Task Teams. https://doi.org/10.1596/33674.

## Figure 4: Digital Government Readiness Components

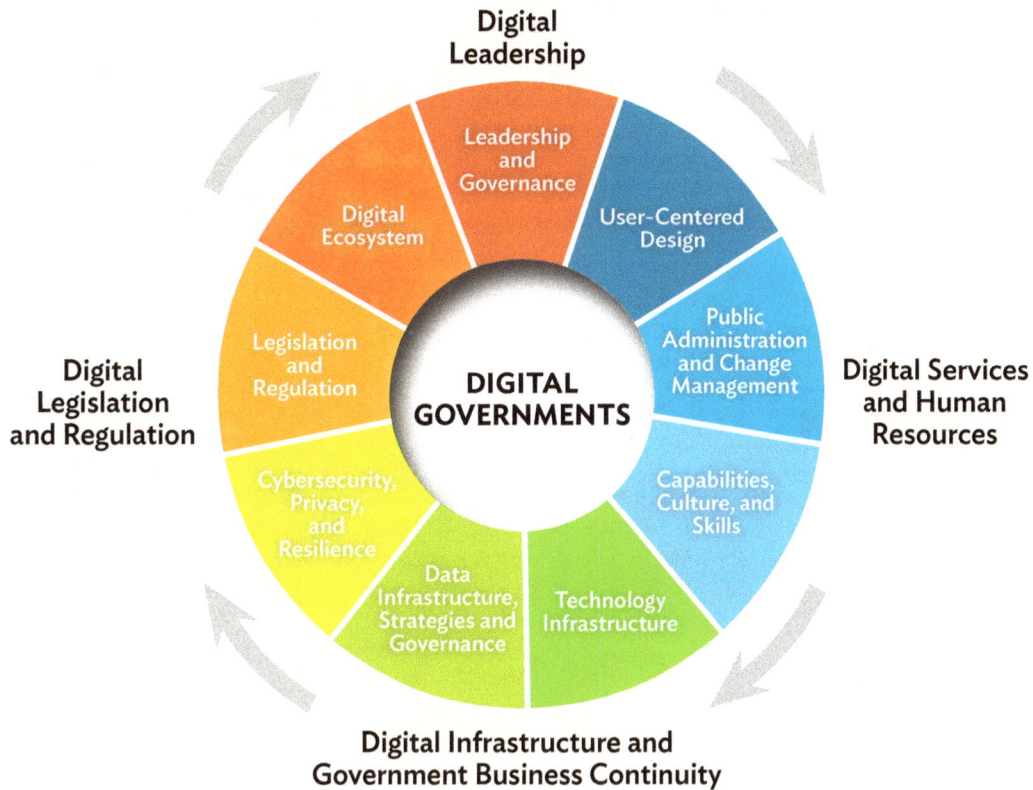

**Digital Leadership**

- Leadership and Governance
- User-Centered Design
- Digital Ecosystem

**Digital Services and Human Resources**

- Public Administration and Change Management
- Capabilities, Culture, and Skills

**Digital Legislation and Regulation**

- Legislation and Regulation

**DIGITAL GOVERNMENTS**

- Cybersecurity, Privacy, and Resilience
- Data Infrastructure, Strategies and Governance
- Technology Infrastructure

**Digital Infrastructure and Government Business Continuity**

### Leadership and Governance
A clear vision, leadership, governance of digital strategy encourages the stakeholders to link the government-wide digital transformation.

### User-Centered Design
Basing high-quality of agile and accessible public services around the users needs—*the public*— increase engagements and open participation of the citizens.

### Public Administration and Change Management
Public administration process has to be optimized for digital delivery. Digital technologies can rapidly improve administrative operations and capabilities.

### Capabilities, Culture, and Skills
Public administration process has to be optimized for digital delivery. Digital technologies can rapidly improve administrative operations and capabilities.

### Technology Infrastructure
Rather than investing specific applications, leaders in today's digital government increasingly look to use whole-of-government standardized technology infrastructure.

### Data Infrastructure, Strategies and Governance
For better decision-making, public spending, and services, digital governments are improving their ability to collect, analyze, and share data using new technologies.

### Cybersecurity, Privacy, and Resilience
A specific protocol, scenarios should be prepared to ensure security and recovery, and minimize risks from any undefined cyber threats, disasters, etc.

### Legislation and Regulation
Legislation and regulation brings transparency to many decision-making on public spending or any e-services that is driven by data.

### Digital Ecosystem
Public digital ecosystem not only boosts innovation, education and entrepreneurship, but also contributes to the modern digital economy.

Source: World Bank. 2020. Digital Government Readiness Assessment Toolkit: Guidelines for Task Teams.

**Table 1:** Using Digital Government Readiness Assessments to Identify Digital Risks

### Pillar 1: Assessing Leadership and Governance

- Does the government have a vision for digital transformation? Is this strategy linked to the national development agenda?
- Are there specific, measurable, and achievable goals toward digital transformation across the various sectors?
- Is there a clear implementation road map that supports the digital government strategy?
- Is there a permanent government entity that owns, maintains, and coordinates the development and implementation of the digital government strategy?
- Are government ICT procurement procedures being followed for digital government programs?

### Pillar 2: Assessing User-Centered Design

- Are there guiding principles established to define the design and implementation of digital or e-services?
- Are users invited to participate in the design, testing, and use of new digital or e-services?
- Is there an integrated multi-channel approach to deliver and promote digital or e-services?
- Are government digital or e-services made accessible to all, taking account of location, connectivity, gender, skills, affordability, and disabilities?

### Pillar 3: Assessing Public Administration and Change Management

- Is cross-government reference data (e.g., personal ID, business registry, land database, etc.) consistently shared electronically across agencies?
- Is government using any management information systems (such as: Integrated Financial Management Information System, e-procurement, HR MIS, Tax MIS, Trade Facilitation, e-business, Education MIS, Health MIS, Land management MIS, Transport MIS, or others)?
- Does the government invest in change management practices toward digital transformation?

### Pillar 4: Assessing Capabilities, Culture, and Skills

- Is there a clear view on the digital capabilities requirements across government to support the digital transformation agenda? Does the government have enough skilled, qualified staff to deliver on the digital transformation agenda? Is the digital transformation strategy clearly understood and communicated?
- Is the government open to outsourcing digital government enabling functions to the local private sector?
- Is there a culture of collaboration around themes or projects among civil service staff in the government?

### Pillar 5: Assessing Technology Infrastructure

- Has a whole-of-government enterprise architecture (covering infrastructure, data, integration, application, operations, and security dimensions) been developed for the digital government program?
- Has the government developed an e-government interoperability framework with mandatory standards for each government entity's systems?
- Has the government designed and deployed an enterprise service bus construct for integrating various data sources to the many service applications?
- Has the government designed and deployed a secure government-wide digital network that connects all entities (at the national and local levels) to share services and data through a secure data center hub?
- Does the government use technologies such as cloud services, IoT, blockchain, or AI–or is it open to the idea of doing so?

### Pillar 6: Assessing Data Infrastructure, Strategies and Governance

- Does the government have a data management strategy (collection, storage, sharing, and re-use)?
- Has the government defined, digitalized, and shared a set of basic data registers? Are all government entities legally required to use this data register?
- Does the government have a data sharing agreement or data exchange protocol with any third party?
- Is government using AI, Big Data, and analytics for decision-making?

### Pillar 7: Assessing Cybersecurity, Privacy, and Resilience

- Has the government developed a cybersecurity strategy and policy?
- Has the government established a cybersecurity unit within a core government entity to manage and maintain security of all digital assets and platforms?
- Is there a computer emergency response team capability in the government? Does the government collaborate with other governments and regional and international organizations to share information on and mitigate cyber threats or risks?
- Does the government have a national critical infrastructure protection plan?

*continued on next page*

**Table 1** *continued*

### Pillar 8: Assessing Legislation and Regulation

- Has a data protection law been enacted that includes minimum elements such as purpose limitation and data minimization, data accuracy, storage limitation, security, transfer of personal data, accountability, and review?
- Has a digital transaction or e-commerce law been enacted?
- Has digital identification legislation been passed that is universal; nondiscriminatory; assigns a unique, random ID number; mandates technological neutrality and interoperability between databases; and provides safeguards regarding data protection?
- Has consumer protection legislation been enhanced to cover e-commerce and e-payments?
- Has a cybercrime law been enacted that criminalizes unauthorized access to computer system, unauthorized monitoring of data, unauthorized alteration of data, content-related offenses, financial crimes, and cyberstalking?
- Has the government passed legislation to support "open access" to government information?

### Pillar 9: Assessing Digital Ecosystem

- Are there innovation hubs and start-up accelerator programs to promote and support innovation?
- Are there established training institutes that offer courses on leading business and technology topics (e.g., agile services, cloud, IoT, AI, data analytics)?
- Have partnerships been formalized with local or international private sector operators in support of digital government?

AI = artificial intelligence, ICT = information and communication technology IoT = Internet of Things, IT = information technology, HR = human resources, MIS = management information system.
Source: World Bank. 2020. Digital Government Readiness Assessment Toolkit: Guidelines for Task Teams.

The tool kit nudges users to identify emerging good practices for digital government and adapt them to their local context, for instance by illustrating how to

- adopt a whole-of-government approach;
- apply a user-centric design philosophy to service design and implementation;
- develop an integrated multi-channel delivery approach;
- design services for scale and sustainability;
- promote collaboration and shared infrastructures, platforms, and processes;
- build capabilities, skills, and cultures for innovation and continuous improvement;
- address cybersecurity, privacy, and resilience; and
- adopt legislation and regulation for addressing digital risks.

The DGRA tool kit has been used for strategic policy dialogues with governments in about a dozen countries (e.g., Cambodia, the Kyrgyz Republic , Uzbekistan, and Viet Nam).[13] In late 2019, the DGRA team launched a Microsoft Excel-based online version of the tool kit to serve as a database for countries' relative digital readiness. Complementing this tool kit, the recently released GovTech Maturity Index measures key aspects of digital transformation in the public sector for 198 economies and assists development practitioners in the design of digital transformation programs.[14] An online data set provides information on the status digital government systems and services.[15]

---

[13] World Bank and Government of Viet Nam. 2019. *Digital Government and Open Data Readiness Assessment*. Washington, DC. https://openknowledge.worldbank.org/handle/10986/32547

[14] C. Dener et al. 2021. GovTech Maturity Index: The State of Public Sector Digital Transformation. Washington, DC. https://openknowledge.worldbank.org/handle/10986/36233.

[15] World Bank. GovTech Dataset. https://datacatalog.worldbank.org/search/dataset/0037889/GovTech-Dataset.

# 1.4. ADB's Country Digital Risk Assessment

Effective tools to assess digital readiness and the compilation of comprehensive risk profiles at country level will help improve project design, make implementation more effective, and allow for more meaningful policy engagements and technical assistance programs by development partners. ADB is considering how to best assess technology risk landscapes and has developed the following country digital risk scan (Table 2) and a country digital risk diagnostic to inform policy and program support for developing member countries (DMCs).

The country digital risk scan attempts to present an at-a-glance view of a country's digital readiness as well as key aspects of its policy, regulatory, and institutional landscape that impact the management of digital risks. The goal is to provide a quick orientation for internal audiences, but is not meant to be exhaustive in its coverage or serve as a cross-country ranking.

The country digital risk scan is organized into ten sections (*marked in yellow*) that cover key aspects of a country's digital risk profile: (i) Digital Strategy for Development, (ii) Policy Design and Technology Selection, (iii) Cybersecurity, (iv) Third-Party Risk Management, (v) Inclusion and Digital Human Rights, (vi) Data Privacy, (vii) Data Management, (viii) Automation, (ix) Environmental Sustainability, and (x) Digital Resilience.

Each section draws on the experience gained through applications of the three digital assessment frameworks (featured in the first part of this document) as well as recently published open-source information, including the following:

- Survey data related to digital policy and regulatory measures (**marked in black**) from ITU's recently launched compendium of "fifth-generation collaborative regulation (G5 Benchmark)."[16] The G5 Benchmark measures the evolution of regulatory and policy frameworks and is meant to help countries navigate the era of digital transformation.[17] It factors in the institutional set-up (agencies and their mandates), policy design principles such as consumer protection and data privacy, regulatory tools to incentivize a sustainable digital ecosystem, as well as policies to promote the digital economy such as innovation frameworks and international partnerships. The International Telecommunications Union (ITU) 5G database provides a ready source of relevant information on digital development. The ADB digital risk scan template draws heavily on its parameters, considering it a good starting point for the scan. Survey data lifted directly from this and other databases should be validated during the scanning exercise
- A series of cross-country indicators, compiled by various international organizations, universities, and policy institutes (*marked in blue*), provides information on global innovation, network readiness, supply chain risk, property vulnerabilities and natural hazard risks, data protection, e-government readiness, national cybersecurity, readiness for frontier technologies, government AI readiness, and internet freedom.

---

[16]  The Country Digital Risk Summary table includes 30 of the 70 indicators used to construct the G5 Benchmark. The 2021 edition is based on self-reported information collected from 193 countries via official ITU surveys, data sets compiled by international organizations, as well as desk research based on official government sources and direct outreach to regulatory authorities.

[17]  ITU. 2022. Benchmark of Fifth-Generation Collaborative Digital Regulation. p 93.

## Table 2: Country Digital Risk Scan

### 1. Digital Strategy for Development

Is there an overarching digital strategy in place?

Is there a forward-looking competition policy, law, or regulation applied to digital markets?

Is there a holistic innovation policy or one tailored to the ICT and digital sector?

Global Innovation Index (2021; 132 countries)

### 2. Policy Design/Tech Selection

Network Readiness Index (2021; 130 countries)

Is there a formal requirement for regulatory impact assessments prior to regulatory decisions?

Are national policy and regulatory frameworks technology and service neutral?

Are there mechanisms for regulatory experimentation?

Are there regulatory sandboxes for digital financial inclusion?

Do ministries and regulatory agencies conduct rolling policy reviews?

### 3. Cybersecurity

Is there cybersecurity/cybercrime legislation or regulation in place?

Is there a cybersecurity institution in place?

National Cyber Security Index (2022; 160 countries)

### 4. Third-Party Risk Management

Does the digital strategy include multiple sectors of the economy?

Does the digital strategy have mechanisms for implementation and operational objectives?

Are there regulatory incentives targeted at network operators or other digital market players?

Has the country adopted a policy, legislation, or regulation related to cloud computing?

FM-Supply Chain Risk Index (2021; 130 countries)

### 5. Inclusion/Digital Human Rights

Internet Freedom Score (2021; 70 countries; 0–39 not free; 40–69 partly free; 70–100 free)

Is public access to information ensured and fundamental freedoms protected, in accordance with national legislation and international agreements?

Can affected parties request reconsideration or appeal adopted regulations to the relevant administrative agency (all sectors)?

Does universal service/access definition include connectivity for telecenters or schools (primary/secondary/post-secondary)?

### 6. Data Privacy

Are there formal data protection rules (e.g., laws or regulations) in place?

Does the country have a data protection authority?

Is there a digital identify framework in place?

Has the country signed international agreements determining jurisdiction and/or managing cross-border flows on data privacy?

Data Protection Rank (2020; 30 countries)

### 7. Data Management

Is there an e-gov/digital first government/national e-government strategy in place?

E-Gov Development Index (2021; 193 countries)

Has the country adopted a policy, legislation, or regulation related to cloud computing?

Has the country adopted a policy, legislation, or regulation related to smart cities?

Has the country adopted a policy, legislation, or regulation related to e-health?

Has the country adopted a policy, legislation, or regulation related to e-applications or m-applications on education and learning?

Has the country adopted any policy/legislation/regulation related to e-apps linked to agriculture/science/financial services?

Are there policies and regulations for e-commerce/e-transactions?

Does the country belong to regional integration initiatives with ICT chapters?

### 8. Automation

Readiness for Frontier Technologies Index (2021; 158 countries)

Has the country adopted a national strategy, policy, or initiative focusing on AI?

Government AI Readiness Index (2021; 160 countries)

Does the country's digital framework include a strategy, policy, or initiative focusing on IoT? Or applied any measure regarding spectrum management and availability for IoT?

### 9. Environmental Sustainability

Has the country adopted e-waste regulations or e-waste management standards?

Are there policy instruments supporting a shift to sustainable consumption and production?

Risk Quality/Natural Hazard Index (2021; 130 countries)

### 10. Resilience

Does the country have a computer security incidence response team in place?

Does a national emergency telecommunications plan exist?

AI = artificial intelligence, ICT = information and communication technology, IoT = Internet of Things.
Source: Authors' compilation.

- The country digital risk diagnostic offers a snapshot view of a country's digital development and underlying risk profile. An initial pilot assessment was carried out for 10 ADB economies to test the approach and the availability of data. As a result of this pilot, a template has been refined for use in future assessments. The experience from the pilot showed that data is not always available, particularly for small economies that are not covered by global indices.
- Each country diagnostic consists of six sections, covering (i) strategic context; (ii) key institutions; (iii) legal and regulatory landscape; (iv) risks related to technology deployment, data, and cybersecurity; (v) digital rights and inclusion; and (vi) sustainability and resilience considerations.

As governments seek to respond to shifting social norms and expectations (Figure 5), several new initiatives, regulations, and institutional reforms are underway across economies. With significant foreign direct investments in emerging markets, novel business models and markets, competitive dynamics, and digitalization of traditional sectors, digital risks and corresponding needs for policy, regulation, capacity development, and investment will continuously change.

ADB's digital risk assessments can help operational staff keep abreast of the rapidly evolving digital landscape. Periodic updates of these assessments, perhaps in annual intervals, are advisable, particularly for economies with growing digital economies or those that are at inflection points in terms of policy initiatives. Learning from experiences with the abovementioned digital assessment tools, it is key to pay close attention to policy developments, institutional arrangements, stakeholders' shifting expectations, and critical gaps between legal and actual application of rules. This would complement global or regional indicators with a more nuanced country-level perspective.

**Figure 5: A Shift in Focus from Digital to Societal Transformation?**

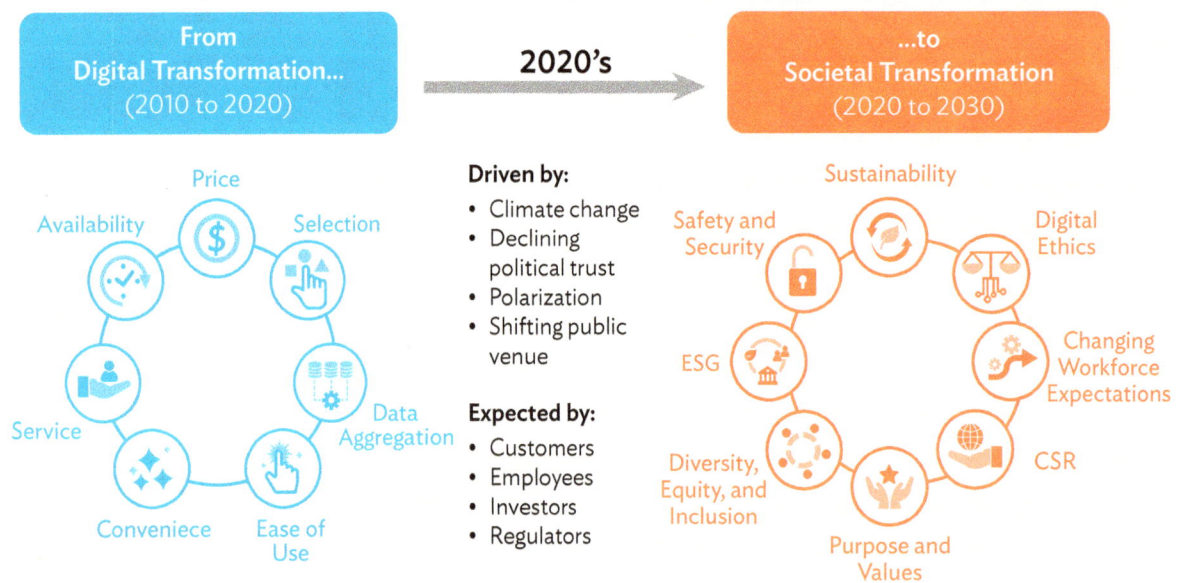

CSR = corporate social responsibility; ESG = environmental, social, and governance.
Source: D. Hebda. 2021. Legal and Compliance Risk Hot Spots for 2022. Gartner (webinar). November.

**Navigating digital challenges.** People exposed to digital risks, highlighting the importance of awareness and cybersecurity (photo by ADB).

# 2

# DIGITAL TRANSFORMATION: CHALLENGES, SELECTION CRITERIA, AND FUTURE TRENDS

The disruptions triggered by COVID-19 underscored the importance of digital transformation. Government institutions and private companies that had invested in digitalization in previous years were able to make rapid adjustments, relying on cloud-based data systems, modern security protocols, an agile organizational culture, and a full range of digitally enabled processes to support the sudden switch to work-from-home and the new reality of virtual transactions and interactions with partners, vendors, citizens or businesses. By contrast, those lagging in their digital initiatives struggled to adjust to the disruption and the rapid economic and social changes that were triggered by the COVID-19 pandemic.

Digital transformation (DX) involves the fundamental rewiring of how services are delivered and how organizations—public and private—operate. By continuously deploying digital technologies, the goal is to build a competitive advantage to improve the quality and/or lower the cost of services being offered.

As the push for DX accelerates globally, applying a systematic process that effectively assesses readiness of countries and enterprises and establishes sound criteria for selecting technologies and vendors that are appropriate for a given context will be important for success. The complex interplay between government institutions, the private sector, and the public will play a key role in the pace and success of digital transformations.

Chapter 2 provides an overview of digital transformation challenges, proposes a set of criteria to guide the selection of digital technologies, and discusses important trends, which help frame the discussion of digital risks in Chapter 3.

## 2.1. Digital Transformation Challenges

Global spending on the digital transformation of business practices, products, and organizations is projected to reach $3.4 trillion in 2026—a rapid growth trajectory.[18]

The Asia and Pacific region is set to become the new hot spot for digital transformation, capturing the largest market share. The use of advanced technologies such as cloud computing, AI, Big Data and analytics, mobility and social media, cybersecurity, and the Internet of Things (IoT), is driving innovation in the region, bolstered by government initiatives and vendor investments, with many governments pursuing tie-ups with major technology companies to rapidly advance the region's digital transformation.[19]

Market pressures and the dynamic development of information technologies are the leading drivers for the rapidly expanding digital transformation market. According to Gartner's 2022 survey of board of directors, nearly 60% rank digital technology initiatives as their top business priority, with more than 70% planning to align risk, strategy, and performance to improve digital resilience.[20]

---

18  IDC. 2022. IDC Spending Guide Sees Worldwide Digital Transformation Investments Reaching $3.4 Trillion in 2026. Press release. 26 November. https://www.idc.com/getdoc.jsp?containerId=prUS49797222.

19  Mordor Intelligence. Digital Transformation Market Size, Share (2022–2027): Industry Trends. https://www.mordorintelligence.com/industry-reports/digital-transformation-market. For instance, the Republic of Korea announced the launch of the Korean New Deal in July 2021 to put digital capabilities and green investments in all industrial ventures.

20  P. Iyengar. *Roadmap to Renewal: Insights from the 2022 Board of Directors Survey*. Gartner webinar on demand.

Private sector organizations tend to allocate their DX investments in three priority areas that align with their digital mission. Many initiatives center on operational objectives, including back-office support and infrastructure for core business functions such as accounting and finance, human resources, legal, security and risk, and enterprise IT. A second set of priorities is focused on innovating and scaling large-scale digital operations, including research and engineering, manufacturing operations, and supply chain optimization. Manufacturing industries are expected to account for nearly 30% of worldwide DX spending in 2022, followed by professional services and retail industries. Meanwhile, the financial services sector will deliver the fastest DX spending growth over the 2021–2025 period.[21] Finally, reshaping the customer experience is a high-growth area, covering all customer-related functions and related technologies supported by DX.

These trends underscore the unparalleled expansion of digital transformation worldwide. Nonetheless, the implementation on the ground paints a decidedly mixed picture: 91% of organizations have adopted or are planning to adopt a digital-first business strategy, but only one in five companies expressed confidence that they have completed their digital transformation journey.[22] According to research by McKinsey[23] and Boston Consulting Group,[24] 70% to 85% of digital transformation initiatives fall short of meeting their business objectives. Even so, the private sector's digital transformation is outpacing the public sector's DX efforts, where success is held back by outdated technologies and a lack of digital and technology skills. According to a survey of 1,200 government officials spanning over 70 countries, nearly 70% stated that their digital capabilities lagged the private sector.[25]

Surveys and practitioner accounts point to a series of challenges and failures that organizations have encountered in achieving their DX outcomes:[26]

- lack of alignment to expected business outcomes, which undermines leadership support for investing in a technology overhaul;
- absence of a digital transformation strategy to transition from legacy systems and applications;
- lack of a change management strategy to raise digital awareness and address internal resistance;
- missing proofs of concept for emerging technologies;
- limited in-house digital and IT skills, exacerbated by poor cross-functional collaboration;
- lack of control over external vendors;
- limited user training to encourage adoption of new tools and processes;
- data privacy and security concerns;
- regulatory and legislative changes; and
- continuous evolution of customer needs and expectations.

---

21  IDC. 2022. Worldwide Digital Transformation Investments Forecast to Reach $1.8 Trillion in 2022, According to New IDC Spending Guide. Press release. 12 May. https://www.idc.com/getdoc.jsp?containerId=prUS49114722.

22  Gartner. Where and How to Target Your Digital Business Transformation. https://www.gartner.com/en/information-technology/insights/digitalization.

23  McKinsey.2019. Perspectives on Transformation. https://www.mckinsey.com/business-functions/transformation/our-insights/perspectives-on-transformation.

24  P. Forth et al. 2020. Flipping the Odds of Digital Transformation Success. *Boston Consulting Group Global*. 29 October. https://www.bcg.com/publications/2020/increasing-odds-of-success-in-digital-transformation.

25  Deloitte. Digital Government Transformation. https://www.deloitte.com/an/en/Industries/government-public/perspectives/digital-government-transformation.html.

26  L. Olmstead. 2021. 9 Critical Digital Transformation Challenges to Overcome (2022). *Whatfix* (blog). 2 December. https://whatfix.com/blog/digital-transformation-challenges/.; and C. Block. 2022. 12 Reasons Your Digital Transformation Will Fail. *Forbes*. 16 March. https://www.forbes.com/sites/forbescoachescouncil/2022/03/16/12-reasons-your-digital-transformation-will-fail/.

To succeed, organizations need to understand and address these challenges and avoid common pitfalls such as resistance to change, challenges in recruiting or retaining the right talent, and compliance concerns, while selecting the best-fit solution. Leadership needs to support and demonstrate commitment to transformation. When the level of awareness and understanding of digital technologies is limited, the adoption of emerging technologies often falls victim to risk-averse decisions. Mechanisms for experimentation (such as a sandbox environment for prototyping and testing) have shown that they are helpful in understanding impact, risks, and the desirable preconditions for a roll out.

Research points to the importance of a holistic, integrated approach that recognizes that DX is more than just a way to offer additional services and achieve greater competitiveness. In fact, digital transformation not only drives organizational change and technological innovation, it also catalyzes broader institutional, societal, and environmental change across the entire ecosystem (Box 1).

Crucially, organizations that laid the groundwork for a successful digital transformation (Figure 6) tended to be more than twice as successful as those that did not. Rather than treating digital transformations as a sprint with a fixed end point, they are better understood as an evolutionary process geared toward continuous improvements over time. Strong leadership matters for setting priorities, communicating progress toward the goals, engaging potential users so as to encourage uptake, and adapting implementation efforts in response to feedback and opportunities for scale-up.

Governments and public sector institutions can learn about digital transformation from the private sector. But important differences remain:

- Digital transformation in the public sector requires caring for access, usability, and user satisfaction, and comes with the necessity of having to meet the policy and political objectives of the day, however well or poorly defined they may be. Legal requirements and social expectations imply that health care, education, justice, and wider public services should be designed to be easily and widely accessible for as many people as possible, considering language, skill, affordability, or technology challenges.
- The metrics for success for digital transformation in the public sector are inherently more complicated. A private e-commerce company cares about conversion rates and revenue or the tailoring of an algorithm to optimize for a specific outcome. Citizen-centricity, productivity, service delivery efficiency, and cost reductions tend to be the driving forces for public services.
- Unlike in the private sector, where consumers or customers drive demand for digital transformation, the process is reversed in public services: users (i.e., citizens, taxpayers, beneficiaries) are nudged to adopt the transformation changes, irrespective of whether they asked for them or not. In the absence of alternative service providers, citizens and businesses who experience inefficient or low-quality transformation results or a biased outcome, will likely be dissatisfied with government performance, potentially leading to an erosion of trust.[27]

---

[27] Compared with the digital experience on private tech platforms, the user experience for public services tends to be cumbersome: processes and interfaces are automated (robotic process application); beneficiaries are required to create user pages to submit application forms and service requests; approvals, rejections, and appeal processes are embedded in individual web portals; and clear communication and grievance mechanisms may be difficult to navigate or be lacking.

Research on the impacts of digital transformation can be categorized in three different clusters: (a) Digital Business Transformation, (b) Technology as a Driver for Digital Transformation, and (c) Institutional and Societal Impacts.

### Cluster A: Digital Business Transformation

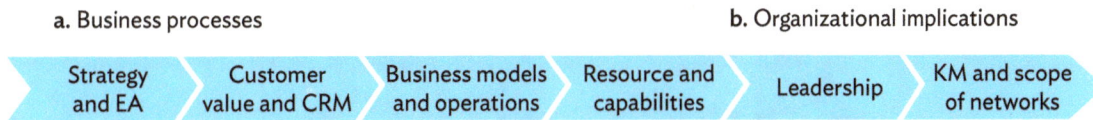

**a. Business processes**    **b. Organizational implications**

| Strategy and EA | Customer value and CRM | Business models and operations | Resource and capabilities | Leadership | KM and scope of networks |
|---|---|---|---|---|---|

### Cluster B: Technology as a Driver for DT

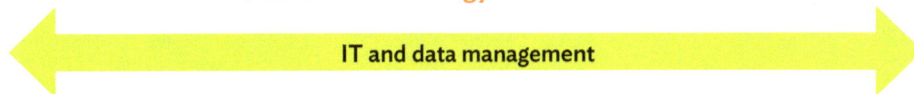

**← IT and data management →**

### Cluster C: Institutional and Societal Impact

**Cluster A:** Most research is focused on changes triggered by DX in business processes and broader organizational implications for leadership, change management, business models, resources and capabilities, customer engagement, and knowledge management. Key issues include the following:

- Leadership: Who is responsible for guiding DX? The DX process calls for strong support from top leadership because DX not only has to be implemented but secured and communicated as well.
- Strategy: Does DX require an alignment of an organization's multiple strategies into one digital business strategy? Alternatively, since content can be distributed through various digital channels, is a stand-alone DX strategy required?
- Agile deployment: In consumer-centric industries, how can enterprises manage the transition from the center (e.g., the enterprise with its supply chains) to the edge (e.g., the digitally active customer)?
- Customer engagement: How can organizations build customer loyalty and trust through innovative, personalized experiences and digitized solutions?
- Business model: The transformation of consumer behavior can change the business models of companies of all sizes. What is required to turn digitalization from a technological issue into a transformational opportunity?
- Knowledge: With the expansion of organizational boundaries and the inclusion of customers, the pooling of knowledge resources has become a key driver of DX. How can open innovation foster an ecosystem where people and organizations can co-create solutions and facilitate the exchange of knowledge, ideas, and technologies with external partners?
- Pathways: DX varies by industry, with banking, retail, and health care more advanced, followed by media and entertainment, while oil and gas are considered latecomers. What shared lessons are emerging?

**Cluster B** covers technology as a driver of DX. How does technology have to be reengineered in the context of rising data relevance? Significantly, research also shows that digital innovations that disrupt industries are not triggered solely by startups, nor does every company need to digitally transform to stay competitive.

**Cluster C** presents the emerging field of institutional and societal implications of DX, focusing on the following questions:

- How do transformational changes disrupt existing organizational structures and values and lead to institutional changes?
- What are the opportunities for science to become more open, collaborative, and global?
- To what extent are existing rules and norms within organizations and networks being replaced, complemented, or threatened by novel actors?
- What sociocultural aspects influence the way innovations gain legitimacy?
- How can economic actors and the environment in which they operate adapt proactively to disruptive changes?

CRM = customer relationship management, DT = digital transformation, DX = digital experience, EA = environmental analysis, IT = information technology, KM = knowledge management.
Source: S. Kraus et al. 2021. Digital Transformation: An Overview of the Current State of the Art of Research. *SAGE Open.* 11 (3).

**Figure 6:** How Do Organizations Lay the Groundwork for a Successful Digital Transformation?

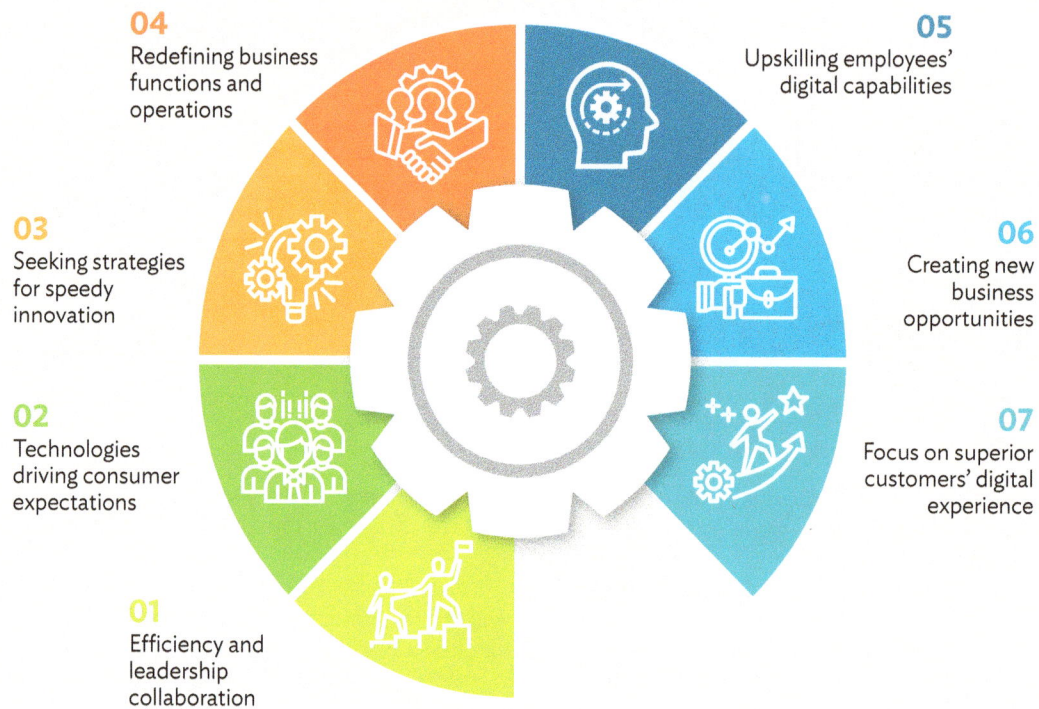

**04**
Redefining business functions and operations

**05**
Upskilling employees' digital capabilities

**03**
Seeking strategies for speedy innovation

**06**
Creating new business opportunities

**02**
Technologies driving consumer expectations

**07**
Focus on superior customers' digital experience

**01**
Efficiency and leadership collaboration

Source: Global Lancers. 2022. Top 7 Reasons Why Digital Transformations Fail.

- The competitive dynamic in the business-to-consumer or business-to-business markets is such that if one business starts to reap the benefits of transformation, new competitors are likely to enter the segment. For government-to-citizen services, there is no alternative: competition is replaced by legal or regulatory provisions. While these are meant to ensure that services are adequate, citizen rights are upheld, and malpractices are avoided, it begs the question which of these two approaches would drive better DX results.

- Finally, the risk calculus in the public sector is different. On the one hand, statutory requirements to provide these services often require fail-safe and redundant systems. On the other hand, when public services fail to deliver, the spotlight of negative impacts and attention is much more widespread compared with transformation failures in the private sector.[28]

For leaders in both public and private sectors, laying the groundwork with careful planning and sufficient foresight, building trust with employees, customers, and the public at large, and treating digital transformations as an evolutionary process, is important.

---

[28] This does not mean that public service organizations should be especially risk-averse; rather, that they should proactively manage risks by using agile approaches, iterating service development, and constantly testing with users.

Addressing the following crosscutting challenges that apply to many, if not all, digital transformation initiatives, can help:

- First, what strategic considerations should guide the selection and acquisition of technologies that are expected to be at the core of new business processes, services, and business models?
- Second, considering the high risk–high-impact trade-offs inherent in most DX initiatives, what criteria can help guide the selection of and engagement with technology vendors?
- Third, given the rapid diffusion of new technologies in both public and private sector domains, how can technology forecasting and foresight provide decision-makers with insights to assess the likely impacts of technologies and help frame plausible scenarios for alternative futures and pathways?

# 2.2. Technology Selection

Digital transformation is important to an organization's life cycle. With so many software applications and hardware solutions available,[29] making the wrong choice can be costly in terms of both time and money.[30] Governments may weigh additional criteria in their decision to select technologies that provide critical digital infrastructure. Public investments in digital public goods, such as digital ID systems, payment platforms, affordable connectivity solutions for remote areas and underserved communities, or data and computing environments for education and health facilities, are often indispensable to achieve public policy objectives such as equitable access, inclusion, and innovation.

Before considering a technology, a first screening needs to determine whether the technology is compliant with applicable standards, laws, and policies. Given the vast scope of possibilities, the selection of technologies should follow a strategic plan and methodical decision-making which focuses on intended outcomes, digital maturity of the organization, maturity of the technology, implementation readiness, trade-offs, and life-cycle aspects (Figure 7).[31]

1. **Validate the need in-depth (the "why").** When considering technology acquisitions, it is critical to align priorities and expected benefits with the organization's vision and business strategy. Most organizations can classify their technology needs into core requirements, nice-to-have requirements, and value added. This distinction is especially important for organizations moving toward a data-centric environment. Various metrics can be used to measure the benefits

---

29   The most common tech solutions for digital transformation include cloud computing; AI, ML, natural language processing; robotic process automation (RPA); API integrations; Big Data, and real-time analytics.

30   CNBC reported that companies like GE, Ford, and Procter & Gamble wasted $900 billion on failed digital transformation initiatives in 2018 alone. The biggest reason for their failure was the inability to effectively communicate their goals, strategy, purpose, and outlook with their employees. K. Kitani. 2019. The $900 Billion Reason GE, Ford and P&G Failed at Digital Transformation. *CNBC*. 30 October. https://www.cnbc.com/2019/10/30/heres-why-ge-fords-digital-transformation-programs-failed-last-year.html.

31   R. Hamzeh et al. 2018. *A Technology Selection Framework for Manufacturing Companies in the Context of Industry 4.0.* Paper prepared for the 2018 World Symposium on Digital Intelligence for Systems and Machines. Kosice. 23–25 August. https://doi.org/10.1109/DISA.2018.8490606. Vera Solutions. 2019. 10 Criteria to Evaluate When Choosing a New Technology. Vera Solutions (blog). 17 October. https://www.verasolutions.org/10-criteria-to-evaluate-when-choosing-a-new-technology/?locale=en.; and C. McKnight. Step-by-Step is the Best Approach to a Successful Technology Selection. *Digital Clarity Group* (blog). http://www.digitalclaritygroup.com/step-step-best-approach-successful-technology-selection/.

of digital initiatives such as improved operational efficiency or workflow automation, effective customer engagements, faster time to market, increased revenue, real-time business analytics, greater innovation, and robust resiliency. A key success factor for clarifying objectives is to define executive sponsor and project team right from the start and to engage with all stakeholders for their input.

2. **Align the DX strategy with the institution's data maturity model (the "what").** Organizations at an infancy level of data maturity face the toughest uphill battle if a full transformation is the end goal. Beyond that level are organizations whose IT departments or specific business lines will be the primary beneficiaries of the new technology. Finally, a fully data-mature organization uses digital transformation technology throughout its operations and can benefit from data insights as an added value stream.

**Figure 7:** 10 Criteria for Technology Selection

Source: Based on Vera Solutions. 2019. 10 Criteria to Evaluate When Choosing a New Technology. With additional insights from authors.

3.  **Determine the technology's intended use (the "what for").** What are the technical requirements of the technology the organization needs? Is it to automate, process, or gather, analyze, and store data? A weighted assessment of multiple potential technologies can balance the requirements—core, nice-to-have, and value added—against their intended use.

4.  **Assess the technology sourcing strategy (the "how").** Determine how much technology development can be done internally. Perhaps a simpler off-the-shelf technology option can more simply and efficiently be added to an existing ("legacy") infrastructure stack to deliver the desired results? Perhaps an open-source solution can be adapted to the identified needs? Before making a technology purchase, conduct a cost–benefit analysis of each option against specific core, nice-to-have, and value added requirements and consider combinations of what are offered; what structures, skills, and expertise exist internally to implement, update, and adapt the solution; and what legacy software or hardware already exists and would need to be updated or retired.

5.  **Weigh the implementation timeline against other options (the "when").** Depending on the chosen pathway, timelines will vary depending on whether to develop the technology internally, merge it with existing technologies, or implement off-the-shelf technology. Alternatives to achieve the same outcome can also affect the sequencing of steps and selection of technologies (Box 2). Another option could be to wait until a more suitable option comes along.

---

**Box 2: How Do Alternative Pathways to Digital Transformation Influence Technology Choices?**

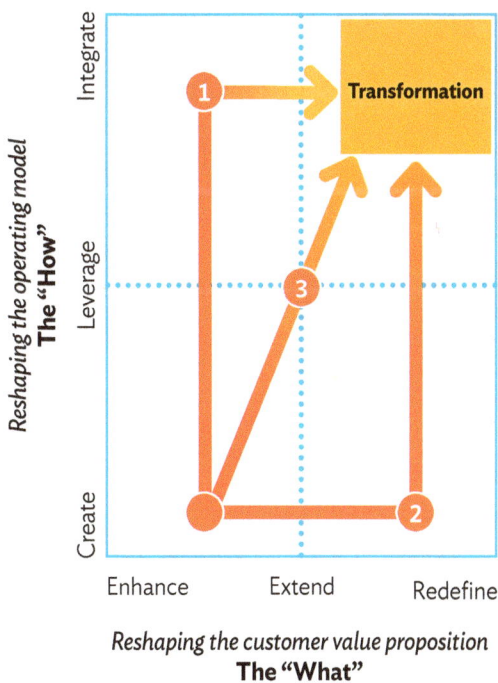

Reshaping the operating model / The "How": Integrate, Leverage, Create

Reshaping the customer value proposition / The "What": Enhance, Extend, Redefine

1.  **Create and integrate digital operations first. Then enhance the customer value proposition based on new processes.** For governments or industries such as manufacturing or mining, where customer expectations are more modest, and the product is either a physical asset or well-established services, changing the way the organization operates is the most suitable strategy for digital transformation. Automating some operations and processes can help reduce costs along with shifting the focus to more value-added tasks.

2.  **Improve customer value proposition with digital content, insight, and engagement, then focus on integrating digital operations.** Industries such as financial services, insurance, and retail, where new revenue-based services can be offered via online and mobile devices, should focus on improving the value they provide to customers.

3.  **Build a new set of capabilities around the transformed customer value proposition and operating model simultaneously.**

**Path 1 technologies:**
- Digital twins
- IoT
- Smart contracts
- Process/Task mining
- Analytics
- Predictive maintenance
- Robotic process automation

**Path 2 technologies:**
- Customer intelligence
- Personalization engine
- Recommendation engine
- Omnichannel
- Chatbot

IoT = Internet of Things.
Source: IBM Institute for Business Value. 2011. Digital Transformation.

6. **Assess the impact of technology investments on the existing IT infrastructure, data migration, cybersecurity, and digital risk profile.** With demand for new digital capabilities continuing to expand, companies and institutions must find ways to manage spending on technology used for running the business, while investing more in technology that grows and improves the business. The most significant cost savings, estimated to range between 30% to 40% (Figure 8), can come from fundamentally rethinking and reengineering the technology and operating model, or by simplifying the services and products alongside the associated IT. Investing in next-generation architecture, moving more workloads onto the cloud (to software-as-a-service and pay-per-use platforms), and advancing automation are all ways to reduce capital expenditure on digital infrastructure and creating life-cycle cost savings.

**Figure 8:** Variable Technology Costs Can Create Room for New Technology Investments

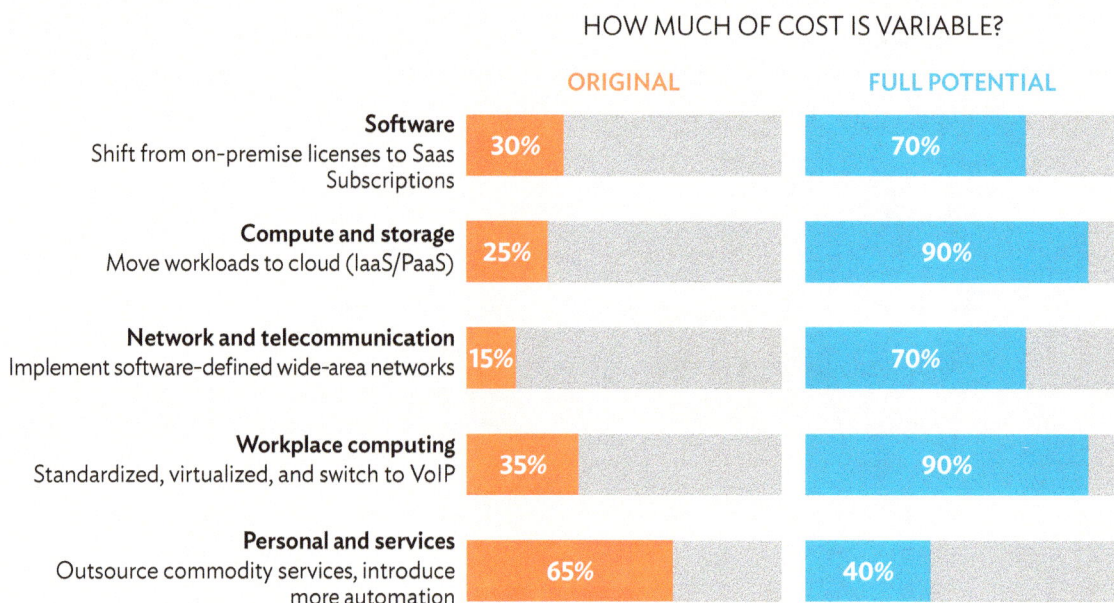

HOW MUCH OF COST IS VARIABLE?

| | ORIGINAL | FULL POTENTIAL |
|---|---|---|
| **Software**<br>Shift from on-premise licenses to Saas Subscriptions | 30% | 70% |
| **Compute and storage**<br>Move workloads to cloud (IaaS/PaaS) | 25% | 90% |
| **Network and telecommunication**<br>Implement software-defined wide-area networks | 15% | 70% |
| **Workplace computing**<br>Standardized, virtualized, and switch to VoIP | 35% | 90% |
| **Personal and services**<br>Outsource commodity services, introduce more automation | 65% | 40% |

IaaS = infrastructure as a service, PaaS = platform as a service, Saas = software as a service, VoIP = voice over internet protocol.
Source: S. Shah, D. Stephenson, and N. Waheed. 2021. Reduce, Replace, Rethink: Transforming Technology Costs. Bain & Co. 20 July.

7. **Consider the fit with the existing technology and data ecosystem ("How good a fit?").** Systems requiring less investment and/or time to implement should be given more weight in the selection process. Data-mature organizations, for example, typically have existing data platforms, while less-mature institutions must first determine their plan for a data platform and whether the selected systems will be interoperable or require adaptations.

8. **Weigh potential trade-offs.** Core functional requirements and the intended data model should carry the most weight during the selection process. In looking for the best return on investment, any trade-off should be viewed through the lens of an organization's strategy and the goals for acquiring the technology and deploying the data infrastructure in the first place, as opposed to the application's features. This is particularly true for those solutions rated as nice-to-have or those who only provide minor value additions.

9.  **Appraise a technology's sustainment and evolution beyond current scope ("Performance over time?").** What types, frequency, and intensity of maintenance and updates are required? Could the new technology's impact be increased by achieving new combinations through extended internal use? Might one need to consider developing additional capabilities? Is the solution's longevity dependent on one vendor, an open-source community, a strong in-house IT capability?

10. **Minimize impact of technological obsolescence ("What's next?").** The problem of technological obsolescence is pervasive and growing. Decision-makers are often faced with a high degree of uncertainty about the rate of technological improvements. A crucial first step is to consider the risks associated with the programs, technologies, and applications being used and to research their life cycle over a 3- to 5-year time horizon. Technologies relating to the internet and the management of enterprise networks are predicted to improve the fastest, including technologies for the delivery of personalized content and software updates over the internet, network security solutions, algorithms, and business process automation. After weighing the advantages and disadvantages of transitioning to a new system, an end-of-life-cycle transition plan with dedicated resources should be developed as the next step.

## 2.3. Vendor Selection

The selection of technology vendors for digital transformation initiatives is a complex decision process that brings inherent risks, but also offers opportunities for mitigating digital risk.[32] Depending on project size and complexity, criteria other than price deserve close attention, such as the nature of vendor involvement, software development processes, relevant experience, and risk management. These include the following:[33]

- ■ **How well does the vendor understand the project's requirements?** Successful companies engage vendors that look at the entire business case and offer comprehensive solutions. The more proactive the vendor team is during the pre-screening and selection process, the higher the chances are that risks and unwelcome surprises can be reduced.

- ■ **Should one choose one or multiple vendors?** The client may decide not to specify all project parameters upfront (during pre-bidding stage) to gain a better understanding of how the vendor proposes to meet objectives by outlining the necessary functionality and clarifying assumptions. A survey conducted by the Boston Consulting Group found that most companies with digital transformation success used advisory firms and a combination of insourcing and outsourcing to execute their programs. This blend of expertise is optimal when the vendor strategy is developed at the start of the initiative.[34]

---

32  V. Duong. 2021. Vendor Selection Criteria: 07 Key Features To Keep In Mind. *Technology Insider*. 18 May. https://www.technologyinsider.asia/digital-transformation/vendor-selection-criteria/.

33  There are resources available online to support vendor selection. Among them the Standardized Information Gathering by the Shared Assessments Group, the Consensus Assessments Initiative Questionnaire by the Cloud Security Alliance, and the Vendor Security Alliance which is an alliance of several large firms.

34  H. Himmelreich and I. Oshri. 2020. Your Digital Transformation Needs a Smart Vendor Strategy. *Boston Consulting Group Global*. 30 October. https://www.bcg.com/publications/2020/your-digital-transformation-needs-smart-vendor-strategy.

■ **What is the vendor's approach to software development?** A vendor's software development life cycle is a cornerstone of the selection process, including how it will be developed, and which models and frameworks the team will use to meet project requirements (e.g., continuous integration, code review, high-end coding review, refactoring, version control). Criteria for quality assurance, bug fixing, user acceptance testing, and protection of user data are best negotiated upfront as part of the vendor selection process. Vendor experience, including past successes and failures with digital transformation projects, should rank high among the selection criteria.

■ **How should vendor relationships and client involvement be reflected as part of contracting and project implementation?** Good relationships and open communication are central to high-performing outsourcing engagements. Successful vendor relationships rely on contracting frameworks that are customized to specific digital transformations. Often, vendors are involved in defining the project scope and implementation plan, with clients shifting their contractual focus from outputs to outcomes. Vendors must also adapt to agile ways of working, which require more frequent interactions and multiple iterations rather than a single final deliverable. Here, continued involvement by the vendor's pre-sale team in the transformation process is crucial in keeping teams working together effectively.

■ **Is the vendor's approach to risk management appropriately tailored to the project and its context?** An important selection criterion is to understand how the vendor will identify and mitigate potential risks connected with project implementation. Possible risks include aspects like third-party service and integration; unforeseen technological challenges (e.g., dependency on tech stack and versioning, changes in user interface design); organizational factors (e.g., change in direction due to client feedback); or added functionality that influences the system's overall performance.

■ **How to keep cost and implementation schedules on track?** Most successful companies reported that staying on budget and delivering by the agreed-upon launch date were key success factors, suggesting a strong link between good planning, competitive selection, and positive outcomes. When a project experiences scope creep or slippage, leaders must be prepared to invoke either contractual obligations for nondelivery or vendor incentives (footnote 34).

Experience shows that most organizations can improve the results of their digital transformation initiatives and manage the unexpected digital risks with the help of a well-thought-out vendor selection and engagement strategy.

This is even more important since significant funding is needed to close the global digital divide by investing in digital infrastructure, applications, and services. Governments for their part have strong incentives to maximize the mobilization of private capital for digital transformation. ITU's *Connecting Humanity* report estimates that $428 billion will be required to connect the 3 billion people ages 10 and above that are unconnected to the internet by 2030, with 68% of that need being concentrated in low- and lower-middle-income countries. The ITU posits an indicative cost sharing for DX investments, with 75% coming from the private sector, largely directed at digital infrastructure, operating expenses, financial services, and digital businesses. The remaining 25% will need to be mobilized from public sources (government budgets, taxes and fees, official development assistance, development bank investments, etc.) to fund digital skills, public platforms, and content, as well as regulation and policy needs. [35]

Multilateral development banks (MDBs) and development partners can play a crucial role in helping to create an enabling environment for private capital flows through a menu of instruments, including challenge funds, pooled procurement, sovereign loans, syndicated bond offerings, public–private partnership project preparation activities, blended finance, investment funds, policy guarantees, and risk transfers.[36]

## 2.4. Technology Forecasting and Foresight

With the future increasingly shaped by technological advancement, the analysis of emerging technologies is receiving increasing attention by governments[37] and corporations.[38] In times of rapid change, decision-makers are rightly concerned about the uncertain and undesirable impact such technologies may have on the economy, society, environment, and culture. In more than a dozen countries (Australia, Canada, European Union [EU] member countries, Finland, France, Germany, Greece, Japan, Singapore, South Africa, the Republic of Korea, United Arab Emirates, the United Kingdom, and the United States), major government agencies[39] and Parliaments[40] have an explicit remit for taking a structured approach to thinking about the future.[41] See also the cross-country perspectives on foresight and anticipatory governance compiled by the Organisation for Economic Co-operation and Development (OECD),[42] the approaches taken by the EU Policy Lab,[43] and the monitoring of emerging technologies and building of future thinking by the World Health Organization.[44] Another widely cited tech trend report with 1 million downloads each year is published by the Future Today Institute.[45]

Technology forecasting attempts to bring potential future technology into focus and predict the rate of technology advancement.[46] It is not meant to provide precise and fully reliable predictions. Risks associated with emerging technologies are hard to assess as their deployment often has not happened yet or has

[35]  ITU. 2020. Connecting Humanity. https://www.itu.int/hub/publication/d-gen-invest-con-2020/.
[36]  Digital Impact Alliance. 2022. Comparative Analysis of Digital Transformation Funding and Financing Models. https://dial.global/research/comparative-analysis-of-digital-transformation-funding-and-financing-models/.
[37]  W. Eggers et al. 2020. How Governments Can Navigate a Disrupted World. *Deloitte Insights*. 24 July. https://www2.deloitte.com/us/en/insights/economy/covid-19/governments-navigating-disruption.html.
[38]  A. Fergnani. 2022. Corporate Foresight: A New Frontier for Strategy and Management. *Academy of Management Perspectives*. 36 (2). pp. 820–844, https://doi.org/10.5465/amp.2018.0178.
[39]  Ross Dawson. 2022. Government Foresight Programs. https://rossdawson.com/futurist/government-foresight/.
[40]  The Office of Technology Assessment at the German Bundestag. About Us. https://www.tab-beim-bundestag.de/english/about-us.php.
[41]  School of International Futures. 2021. *Features of effective systemic foresight in governments around the world*. https://assets.publishing.service.gov.uk/government/uploads/system/uploads/attachment_data/file/985279/effective-systemic-foresight-governments-report.pdf.
[42]  OECD. Foresight and Anticipatory Governance in Practice. https://www.oecd.org/strategic-foresight/ourwork/Foresight_and_Anticipatory_Governance.pdf.
[43]  EU Policy Lab. 2023. Technology Foresight: Anticipating the Innovations of Tomorrow. 26 May. https://policy-lab.ec.europa.eu/news/technology-foresight-anticipating-innovations-tomorrow-2023-05-26_en.
[44]  WHO. 2022. WHO Foresight: Monitoring Emerging Technologies and Building Futures-Thinking. https://www.who.int/activities/who-foresight---monitoring-emerging-technologies-and-building-futures-thinking.
[45]  Future Today Institute. What We See. https://futuretodayinstitute.com/trends/.
[46]  Arthur D. Little. Technology Foresight: Anticipating Future Impact. https://www.adlittle.com/en/insights/viewpoints/technology-foresight-anticipating-future-impact.

not occurred at a sufficient scale to assess intended and unintended impact. Despite these limitations, technology forecasting can generate valuable insights into technology development processes and identify possibilities and scenarios.

Technology forecasting methods are well established and continue to evolve, thanks to increased availability of information and computational capacity. In practice, the combination of several forecasting methods is beneficial because each method can only deal with limited aspects. Available techniques, both qualitative and quantitative methods, fall into two broad categories: [47]

- Exploratory technological forecasting aims to predict futures based on the past and present state of the art. These techniques include Delphi method, expert surveys, environmental scanning, cross-impact analysis, trend extrapolation, and social network analysis, among others.
- Normative forecasting techniques, such as futures thinking and foresight, take a different approach and trace backward from future goals, missions, and needs to determine the necessary actions to achieve these points. This category includes causal layered analytics, back-casting, future triangles, scenario planning, and road mapping, among others.

Germany's Federal Ministry for Education and Research is piloting a hybrid approach, which puts emphasis on the interdisciplinary perspectives of social and technological change by combining its innovation and technical analysis, with a time horizon of up to 5 years, with foresight, thereby extending the timeline of innovations and the associated changes of up to 15 years to include wider social, ecological, economic, and ethical and regulatory aspects.[48]

Many development programs have implementation horizons of 7–10 years, making them particularly susceptible to technological changes, as the initial technology package may become outdated before project completion. The following three examples, drawing on insights from technology entrepreneurs, industry specialists, and scenario exercises, help illustrate different approaches and time horizons.

The first example draws on views from the technology pioneer community of early- to growth-stage companies who were asked by the World Economic Forum in early 2022 about their predictions of technological changes that could change the world over the next 5 years. Their views—which need to be interpreted as signals from the innovator community and potentially leaning in the direction of their preferred future—are summarized in Table 3:

[47]  G. Calleja-Sanz, J. Olivella-Nadal, and F. Solé-Parellada. Technology Forecasting: Recent Trends and New Methods. In C. Machado and J. Paulo Davim, eds. *Research Methodology in Management and Industrial Engineering*. Management and Industrial Engineering (Cham: Springer International Publishing, 2020), 45–69, https://doi.org/10.1007/978-3-030-40896-1_3.
[48]  Bundesministerium für Bildung und Forschung. Insight. https://www.bmbf.de/bmbf/de/forschung/soziale-innovationen-und-zukunftsanalyse/insight/insight.html.

## Table 3: 2022 Predictions by World Economic Forum's Technology Pioneer Community

| Tech Forecast | Application |
|---|---|
| Financial inclusion | With the proliferation of gig work, digital labor platforms will increasingly embed financial services into their products and make credit available to those ignored by traditional financial institutions. (*Vahan*) |
| Digital currency | Central bank digital currency has the potential to revolutionize the financial system by offering offline payments, shielded transfers, automation, and cash-like properties. (*Fluency*) |
| Web3 | By 2025, Web3 technologies will have changed e-commerce by enabling the seamless exchange of digital assets for physical products, services, and experiences through tokenization across platforms and automated settlement by smart contracts. (*Boson Protocol*) |
| Data industry | The growth in data has created issues of privacy, affordability, access, and security. In the coming years, if governments allow it, the data industry could become more inclusive, affordable, and transparent. (*Databento*) |
| Quantum computing | The quantum internet will have a profound impact on how we live our lives by enabling breakthroughs in energy, medicine, data security, remote sensing, material sciences, and other fields. (*Aliro Quantum*) |
| Metaverse | The metaverse will be shaped by the communication of our emotions and enabled by technologies such as VR/AR and brain–computer interfaces. New platforms, disciplines, and senses will redefine social contracts in the virtual world where emotion, trust, and safety become our most important currencies. (*Emerge*) |
| Carbon credit monitoring | Nature-based solutions are key to removing $CO_2$ from the atmosphere and sequestering it. With increased demand for verifiable carbon credits, remote sensing data streams from high-resolution cameras and lasers, combined with cloud data processing, AI, and data fusion, will monitor natural ecosystems. (*Agerpoint*) |
| Sustainable construction | The construction industry is adopting battery Energy Storage Systems (ESSs) for construction sites with the goal to reduce carbon emissions by 80%. Long-duration ESSs include flow batteries, gravity-based ESSs, non-lithium batteries, and hydrogen, which enable offsite recharging from solar or wind farms. (*Ampd Energy*) |
| Smart buildings | Buildings account for 50% of carbon emissions. Smart buildings using sensors, digital twins, and AI will dynamically respond to space occupancy, weather, and more, and adjust to support human wellness. (*Akila*) |
| Smart grid technology | With more intermittent renewable energy in the power mix, a clean energy future requires grid flexibility to always meet demand. By 2025, software solutions that allow smart energy technologies such as EV chargers and heat pumps will respond to real-time grid requirements and optimize asset owner's earning. (*Leap*) |
| Advanced manufacturing and fashion tech | The apparel industry is forced to mass-produce goods with limited consumer input, resulting in high return rates, waste from overproduction, and low margins. Advanced manufacturing and fashion technology can produce custom goods locally, dramatically reducing turnaround times and environmental footprint. (*DXM*) |
| Predictive supply chain technology | Climate change, pandemics, conflicts, and fragmented supply chains have impacted food production and distribution. By 2027, major food and consumer goods manufacturers will use AI-driven supply chain technology to predict changes in food availability and purchasing habits and act before weather, labor issues, and other incidents can harm the global food supply to keep shelves stocked. (*Everstream*) |
| Nourishing food | With new processing methods for healthy sugar or salt alternatives, good food will become an accessible, ubiquitous option for all income groups. People will eat more nourishing food because cost and taste will be on par with less-healthy, legacy food and reduce diet-related health-care costs. (*Bonumose*) |

*continued on next page*

**Table 3** *continued*

| Tech Forecast | Application |
|---|---|
| Education technology | Technological progress can help every child develop skills and competencies to co-create solutions to the world's biggest problems. AI will be used to understand children's own interests to suggest the next step in their learning journey and generate insights for parents and teachers. (*Twin Science*) |
| Talent markets | Workplaces have gone through major changes, warranting a full re-think. The pandemic also changed the talent landscape for many companies to a permanent global and hybrid one. This means that technology will be a true leveler, bringing the best opportunities to the best talent irrespective of where they live. (*Onloop*) |
| Precision medicine | Between now and 2030, over 1 billion people will suffer from infertility. In-vitro fertilization is expensive, physically, and emotionally draining, and relies on complex clinical decisions. By 2027, AI will power clinical decision making in fertility clinics, enabling physicians to deliver a new level of precision medicine (*Alife Health*) |

AI = artificial intelligence, AR = augmented reality, $CO_2$ = carbon dioxide, ESS = energy storage system, EV = electric vehicle, VR = virtual reality.
Source: World Economic Forum. 17 Ways Technology Could Change the World by 2027.

The second example stems from the technology research company Gartner, which publishes annual hype cycles of emerging technologies for industries, government institutions, and policy domains. Intended to represent common patterns that arise with the maturity, adoption, and social application of technologies, they may help inform and reduce investment decisions for technologies, many of which are not expected to become mainstream for another 5 to 10 years, if ever. The five phases in the Gartner Hype Cycle are Technology Trigger, Peak of Inflated Expectations, Trough of Disillusionment, Slope of Enlightenment, and Plateau of Productivity.[49] The 2022 hype cycle for government services, for instance, highlights the potential for improving citizen focus through the adoption of human-centered design principles and the use of superapps for delivering personalized services (Figure 9).

Third, futures thinking and foresight approaches can generate valuable insights over a 20–50 year horizon, where Horizon 1 is the present, Horizon 2 is the uncertain emerging future 5–20 years forward, and Horizon 3 is the desired long-term future 20–50 years forward. Facilitated as highly participatory exercises to strengthen cross-sector links, inspire the emergence of integrated solutions, and empower people to create the future they desire, participants are encouraged to time the future differently.[50]

In reimagining what the future of transport in Asia and the Pacific might look like by 2050, ADB developed a series of eight plausible futures. The overriding goal was that these mobility innovations must be driven by clear outcome-led planning to address specific social, environmental, and economic needs of transport users. The road map for achieving the full integration of transportation and technology by 2030, for instance, requires data analytics, first- and last-mile delivery, safe and secure autonomous system, and multimodal coordination (Figure 10).

---

[49] Critics of the hype cycle note that the outcome does not depend on the nature of the technology itself, that it is not scientific in nature, and that it does not reflect changes over time in the speed at which technology develops. Wikipedia. 2022. Gartner Hype Cycle. 28 August. https://en.wikipedia.org/w/index.php?title=Gartner_hype_cycle&oldid=1107237130.

[50] ADB. 2020. *Futures Thinking in Asia and the Pacific: Why Foresight Matters for Policy Makers.* 1 April. https://doi.org/10.22617/TCS200126-2.

**Figure 9:** Gartner's 2022 Government Hype Cycle: Delivering Services with a Citizen Focus

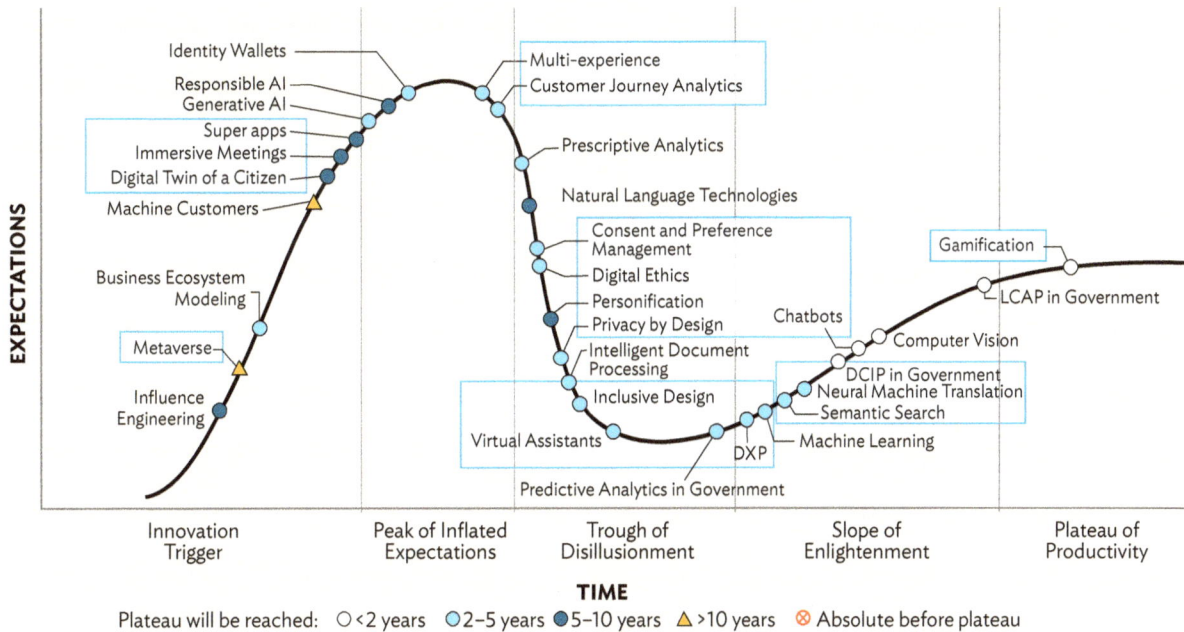

AI = artificial intelligence, DCIP = document-centric identity proofing, LCAP = low-code application platform.
Source: Gartner Webinar. Gartner Hype Cycle for Digital Government Services, 2022, August 2022. https://www.gartner.com/en/webinars/4016921/gartner-hype-cycle-for-digital-government-services-2022.

Development sector spending and related infrastructure investments have long life cycles. The technology solutions that are designed "into" projects need to be informed by an over-the-horizon view as advancements in technologies can quickly overtake the initial project design. Futures thinking and foresight approaches can be helpful for decision-makers who have to plan for the long term and take decisions based on current day knowledge of and experience with technologies that are constantly changing.

**Figure 10:** ADB's Futures Thinking: Principles and Strategies for Technology-Enabled Transport Services by 2030

| Scenarios | Strategy: Baseline | Strategy: Progressive | Strategy: Transformative |
|---|---|---|---|
| **Principle 1**<br>*Data analytics and digital transformation* | **Data Capture**<br>Gather data for key business, asset management, or operational processes using Internet-of-Things sensors. Apply data insights to optimize and resolve bottlenecks in transport networks and improve passenger and freight journeys. | **Data Collaboration**<br>Sharing of data analytics among users and key stakeholders to better manage assets, support decision-making, provide real-time transport journey planning, and streamline business processes for transport service providers (i.e., introducing shared portals). Smart cities concept used to enhance transport user experience and efficiency. | **Digital Transformation**<br>Utilize simulation tools or deploy digital twins to support scenario planning for more resilient transport system operations. |
| **Principle 2**<br>*Facilitate first- and last-mile connectivity for passengers and urban logistics* | **Efficient Transfers**<br>More efficient point-to-point transfer thanks to freight consolidation. Increased use of lockers systems for pick up and delivery. Personalized transport services introduced as part of wider physical and digital journeys. Ensure roads are designed to meet the needs of a variety of transport modes (i.e., walking, micro-mobility, two-three wheelers) | **Flexible Multimodal Transfers**<br>Configurable and fluid transfer across transport modes through a centralized system, enabling both the mobility and delivery network to be more resilient and flexible. | **Predictive Connectivity**<br>Transform systems at service level, technology incorporates human behavior into a system configuration of autonomous vehicles, and drones. Smartphones learn and predict rider preferences and pre-book or suggest a variety of modes for first- and last-mile connectivity. |
| **Principle 3**<br>*Ensure safe and secure autonomous systems* | **Safe Autonomous Systems**<br>Understand the risk of data capture and sharing, privacy issues, trade-offs between security and freedom and learn about technological processes in variable environments to minimize risks and improve security (i.e., less crashes, no hacking the system), and develop data capture and storage plans. | **AI-Enabled Systems**<br>Utilize artificial intelligent (AI) and machine-learning-enabled systems to collect data, detect threats, and mitigate risk and vulnerabilities of autonomous systems. | **Optimized Functionality**<br>Optimize systems to deliver repeatability and efficiencies through automation to continuously improve functionality (i.e., in control centers for route guidance, and maintenance schedules). |
| **Principle 4**<br>*Seamlessly manage and coordinate multimodal networks* | **Dynamic Modal Split**<br>Journeys suggested on mobility apps that include dynamic modal split and consider time of day, congestion, and popularity to influence route choice. | **Flexible Network Management**<br>Responsive and flexible network and system able to manage and respond to different peak/off-peak times. | **Mode-Agnostic Payment**<br>Travel-as-a-service is implemented citywide, enabling transport users to choose and pay for travel by distance and time, regardless of transport mode. |

Source: ADB. 2022. *Reimagining the Future of Transport Across Asia and the Pacific*. Manila.

**Unveiling the future.** Embracing digital transformation as economies evolve, innovate, and adapt to ever-changing landscape (photo by ADB).

# 3

## DIGITAL RISK

Many government institutions and companies are under pressure to scale digitally, be it to strengthen resilience, provide greater value, deliver on rising expectations for better access to services, or pursue new avenues for innovation and growth. These priorities are borne out in the most recent digital landscapes for 2022–2023.

For governments, the trends shaping the post-pandemic period can be captured under three headlines:

(i) Building long-term resilience to anticipate and adapt to future shocks, such as technology shifts, climate change, geopolitical upheaval, widening geopolitical digital divide and rivalries, supply chain shocks, or other disruptions.

(ii) Overcoming interagency boundaries that have limited governments' ability to address complex social problems by overhauling and integrating government structures, systems, and data-sharing arrangements, as well as acting as a catalyst in the innovation ecosystem.

(iii) Making public programs and social services more accessible, responsive, inclusive, and truly equitable.[51]

For companies, most boards flagged digitalization as the single biggest strategic business priority (58%), followed by pressures to retain and recruit the workforce needed for digital acceleration (52%), and the (belated) recognition that environmental, social, and governance (ESG) issues present an emerging risk that requires changes to corporate values (32%). Most company boards plan to raise their organizations' risk appetite (57%)—a shift that implies an expanded role for risk management and the elevation of cybersecurity risk as a top priority.[52] As a result, eight in 10 corporate boards see improved risk management as critical for their business to protect and build value in the next 5 years.[53]

The similarities between these public and private sector perspectives are striking:

- Climate resilience has risen to the top of the agenda for leaders in government and corporations.
- Efforts to diversify supply chains have become increasingly urgent to ensure business continuity and economic resilience.
- A widening geopolitical divide, in which lagging countries become even more dependent on digital technology providers in advanced economies, may prevent them from competing effectively as opportunities have shifted in favor of digitally advanced economies and elites—a trend that is set to continue.
- Future-proofing the labor force will require a fundamental rethink of education, skills training, and employment frameworks.
- Rising expectations for integrated, seamless service delivery challenges leaders to dismantle organizational silos and find new ways to connect with citizens, clients, and consumers.

---

[51] Deloitte Insights. Government Trends 2022. https://www2.deloitte.com/us/en/insights/industry/public-sector/government-trends.html.

[52] The 2022 Gartner Board of Directors survey was conducted in mid-2021 among 273 respondents in the US, Europe, and Asia and the Pacific. K. Panetta. 2021. 6 Key Takeaways from the Gartner Board of Directors Survey *Gartner*. 21 October. https://www.gartner.com/en/articles/6-key-takeaways-from-the-gartner-board-of-directors-survey.

[53] T. Dekker. 2021. The Board Imperative: How Can Data and Tech Turn Risk into Confidence? *Ernst & Young*. 14 July. https://www.ey.com/en_gl/risk/how-can-data-and-tech-turn-risk-into-confidence.

- The push toward data collaboration requires investments in cloud-based systems and advanced data capabilities.
- Remote work, telehealth, and virtual instructions have spurred digital adoption during the COVID-19 pandemic, but also revealed persistent digital divide and equity issues that could leave billions outside of digital life.
- People's trust deficits are mounting over data breaches, unresolved ethical concerns, and privacy violations.

Despite strong upsides, the conundrum facing governments and companies is this: the greater the digital transformation, the greater the associated digital risks. Digital investment will continue to grow and so will the associated digital risks. The drivers of digital risks are not only products, services, hardware and software, but also their increasing interconnectedness, and convergence resulting in unintended consequences and unforeseen risks.[54] The key is to identify, mitigate, and monitor risks to minimize their impact proactively and continuously.

# 3.1. The Spectrum of Digital Risks

Digital risks have emerged as one of the fastest-growing, most pervasive risk categories for companies, governments, and even entire countries that can affect organizations of any type, size, sector, and maturity.

Digital risks can be defined as the risks associated with the creation, delivery, and use of new digital technologies, processes, and services that are deployed to achieve operational efficiencies, scale new business models, or deliver new services to customers or citizens.[55] This combination of technologies, data, business processes, and strategic objectives can have unwanted and, at times, unforeseen consequences that can undermine digital transformation efforts and thwart the achievement of intended outcomes.

Digital risks, including cyber, third-party, business continuity, legal and regulatory, conduct, ethical, or data privacy risk, create uncertainties and may cause financial and reputational losses. For public sector agencies, digital risks can involve the inability to deliver services; the erosion of trust in government, science, and social cohesion; a loss competitiveness and international credibility; and/or declining government revenues (e.g., a month-long cyber attack paralyzed Costa Rica's tax and custom agencies). That said, without engaging in digital transformation programs in the first place and being cognizant of the attendant risks, governments and enterprises would forgo the prospect of future growth, revenue generation, service delivery, and competitiveness.[56]

[54] J. Wheeler. 2022. Accelerated Digital Investment Demands Integrated Risk Management. *AuditBoard*. 12 October. https://www.auditboard.com/blog/accelerated-digital-risk-investment-demands-integrated-risk-management/.

[55] CIO Summits. Managing Digital Risk: 8 Types of Digital Risk and How to Manage Them. https://www.ciosummits.com/Online_Assets_RSA_How_to_Manage_Eight_Types_of_Digital_Risk.pdf.

[56] AuditBoard. 2022. Digital Risk Maturity Report 2022: Turning Digital Risk Into Your Competitive Advantage. https://www.auditboard.com/resources/ebook/digital-risk-maturity-report-2022-turning-digital-risk-into-your-competitive-advantage/.

To appreciate why digital risks pose such a complex set of challenges to public and private organizations, consider the multiple ways in which digital transformation initiatives can be derailed:

- The launch of new business models (or service delivery channels, if translated into a public sector context) may be impeded by third-party vulnerabilities that could result in supply chain disruptions; regulatory gaps and barriers that undercut their viability (e.g., ride sharing, electric vehicles); negligent compliance efforts with newly adopted laws and regulations; or process automation risks.
- The roll out of new services may fail due to lack of adoption as a result of poor digital literacy, lack of affordability, or shortage of relevant local (language) content.
- Digital technologies may cause or contribute to adverse and unforeseen outcomes that enable the erosion of civil rights (e.g., digital surveillance), deepen digital exclusion for vulnerable groups, or usher in new workplace safety risks.
- Efforts to improve operational efficiencies may be foiled due to cyber attacks that exploit the growing attack surface for data breaches during a digital transformation process, accidental exposures of sensitive data, risks arising from the deployment of new cloud solutions or changes in data architecture, or operational risks caused by digital skill gaps.

Broadly speaking, digital risks occur in four main domains (Figure 11): Security, Soundness, Sustainability, and Safety.

## Figure 11: Digital Risk Domains

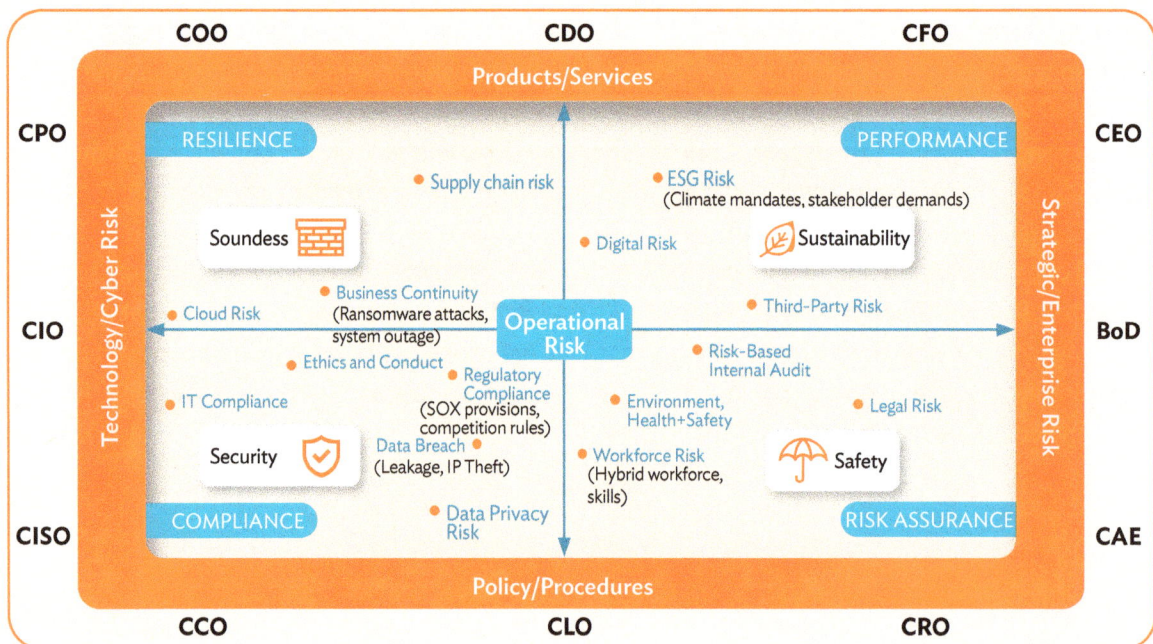

CEO = Chief Executive Officer; BoD = Board of Directors; CAE = Chief Audit Executive; CRO = Resource Officer; CLO = Chief Legal Officer; COO = Chief Compliance Officer; CISO = Chief Information Security Officer; CIO = Chief Information Officer; CPO = Chief Product Officer; COO = Chief Operating Officer; CDO = Chief Data Officer; CFO = Chief Financial Officer.
Source: Based on J Wheeler. 2022. 2022: The Year of Digital Risk Discovery. *AuditBoard* (blog). 26 January. With authors' modifications.

## (1) Security Domain

- IT compliance risks cover rule violations related to the installation and operation of digital systems in companies and public organizations. These rules usually determine requirements for IT security, data protection, data availability, and data integrity that an organization must comply with to meet legal standards or contractual agreements with customers and business partners.[57] Violations of IT compliance can be subject to substantial fines.

- Regulatory compliance risks involve threats posed to a company's financial, organizational, or reputational standing resulting from violations of laws, regulations, codes of conduct, or organizational practices. Compliance is a particular concern in countries with data and privacy laws (e.g., the EU General Data Protection Regulation, California Consumer Privacy Act), for regulation-intensive sectors such as health care or finance, or for clients with high confidentiality requirements. The consequences of failing to meet compliance obligations can include regulatory fines, civil litigation from third parties, debarment from bidding on future contracts, or loss of a company's reputation with would-be customers.[58]

- A data breach is a security incident in which malicious insiders or external attackers gain unauthorized access to confidential data or sensitive information such as financial information, matters of national security, trade secrets, medical records, or personally identifiable information. Data breaches can occur through accidental data leaks, malware attacks, social engineering, lost or stolen hardware, or lack of access controls.[59]

- Data privacy risks address the proper handling, processing, storage, and usage of personal information. At the center are the rights of individuals and/or community groups with respect to their personal information. The most common concerns regarding data privacy involve confidentiality in contract management, the application of regulations or laws, or management by third parties. Any risk assessment is performed from the perspective of protecting the rights and freedoms of individuals and/or communities that could be adversely affected.[60]

- Ethics and conduct risks relate to failures in ethical responsibilities (as well as legal obligations) to protect customers' data, defend against breaches, prevent biases, or ensure that personal data is not compromised.[61] For instance, the potential of AI in health care cannot be fully realized unless major ethical issues are addressed, including informed consent, safety and transparency, algorithmic fairness, and data privacy.

## (2) Soundness Domain

- Supply chain risks arise from exposures, threats, vulnerabilities, and disruptions posed by suppliers, their respective suppliers, and the products or services they provide.[62]

---

57  Myra Security. What Is IT Compliance? https://www.myrasecurity.com/en/what-is-it-compliance/.

58  Reciprocity. What Is Regulatory Compliance? https://reciprocity.com/resources/what-is-regulatory-compliance/.

59  Tech Target. What Is a Data Breach?. https://www.techtarget.com/searchsecurity/definition/data-breach.

60  Data Privacy Manager. Data Privacy vs. Data Security (definitions and comparisons). https://dataprivacymanager.net/security-vs-privacy/.

61  Efforts are underway to embed ethical principles such as transparency, accountability, and explainability into the creation of products, tools, and services to build public trust and confidence in technology.

62  Joint Task Force Interagency Working Group. 2020. Security and Privacy Controls for Information Systems and Organizations. *NIST Special Publication 800-53 Revision 5*. Gaithersburg, MD: National Institute of Standards and Technology. https://doi.org/10.6028/NIST.SP.800-53r5.

- Business continuity risks include threats that disrupt the functioning of a business or the ability of an organization to maintain the regular or planned delivery of products or services following a disruptive incident.[63] Disruptions can be caused by ransomware attacks, system outages, disasters, and other force majeure events.
- Risks specific to cloud computing can occur when data is shifted from an on-premise server infrastructure to cloud infrastructure. These also are related to changes in data architecture, the adoption of complex private, public or hybrid cloud models, and/or the deployment of new digital business operations or IT systems. Examples include cloud server outages, disruptions due to misconfigured servers, unauthorized access, as well as vendor lock-in.

## (3) Sustainability Domain

- ESG risks refer to adverse climate change impacts; failed mitigation and adaptation efforts; insufficient environmental management practices; inappropriate working and safety conditions; negative social impact of digital services and products; and a lack of accessibility and inclusion, respect for human rights, anti-bribery and corruption practices, community relations, or compliance with relevant laws and regulations.[64]
- Third-party risk refers to the likelihood that an organization will experience an adverse event (e.g., data breach, operational disruption, reputational damage, financial loss, or strategic planning failure) through service or product provided by third parties. Third parties can include software vendors, suppliers, staffing agencies, consultants, financial partners, and contractors which are typically engaged in DX efforts.
- The digital transformation effort may also fail to achieve and/or sustain the expected outcomes.

## (4) Safety Domain

- Environment, health, and safety risks are subject to laws, regulations, and workplace efforts to protect the health and safety of employees and the public as well as the environment from injuries, illnesses, and harmful environmental releases.[65]
- Workforce risks may affect an organization's ability to recruit new talent and/or retain employees who are experts in their field. Today's flexible workforce and hybrid work environments may require changes to the organization's value proposition to meet employees' expectations. Insider threats, posed by employees with access to sensitive information, are another, often overlooked risk.[66]
- Legal risks may concern risks of losses or damages arising from unintentional or negligent failure to comply with statutory or regulatory safety obligations, or from the nature or design of a product or service.

---

[63]   VComply Editorial Team. 2020. What Is Business Continuity Risk? *VComply* (blog). 4 December. https://www.v-comply.com/blog/what-is-business-continuity-risk/.

[64]   A. Gorley. 2022. What Is ESG and Why It's Important for Risk Management. Sustainalytics (blog). 2 March. https://www.sustainalytics.com/esg-research/resource/corporate-esg-blog/what-is-esg-why-important-risk-management.

[65]   Digital environment, health, and safety initiatives include virtual reality technology to improve training effectiveness; drones and robotics to perform dangerous, dirty, and difficult jobs; AI bots to automate complex tasks; workplace sensors for performance measurement; or biometrics to gather physical and mental health data. R. Dabbs. 2019. How Can Your Digital Strategy Help Improve EHS Outcomes? *Ernst & Young*. 12 August. https://www.ey.com/en_us/assurance/how-can-your-digital-strategy-help-improve-ehs-outcomes.

[66]   RiskOptics (formerly Reciprocity). 2022. 10 Common Types of Digital Risks. *RiskOptics* (blog). 16 June. https://reciprocity.com/blog/common-types-of-digital-risks/.

■ Quality risks may be caused by implementation problems and the potential for losses due to subpar product or service quality.

Surveys of board members of global corporations offer a snapshot how these risk categories vary across sectors and geographies (Figure 12) as well as how rapidly changes in circumstances may trigger an adjustment in risk posture.

**Figure 12: Comparing Risks across Industries and Regions, 2021–2022**

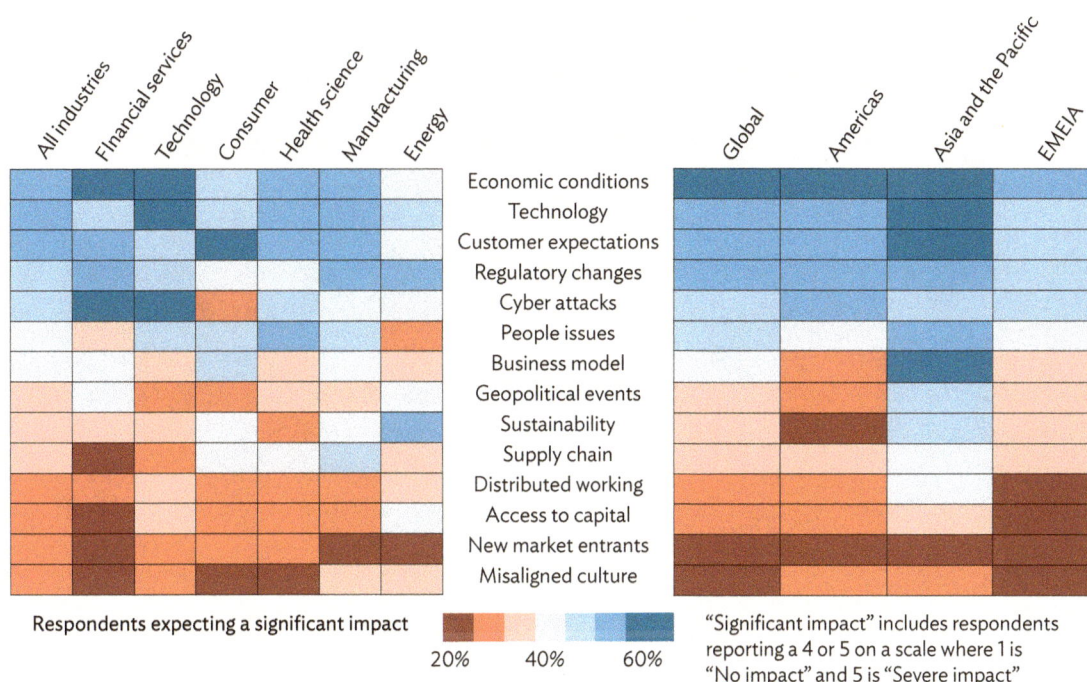

EMEA = Europe, the Middle East, and Africa.
Source: S. Sutherland, T. Dekker, and K. Pederson. 2021. The Board Imperative: Is Now the Time to Reframe Risk as Opportunity? *Ernst & Young.* 14 July.

Queried in mid-2021, company boards across all industries ranked economic uncertainty, technology, consumer expectations, and regulatory changes as the most significant sources of risk.[67] Cyber attacks and data breaches were ranked as the fifth most important risk across all sectors, but were rated first by executives in financial services, technology, media and entertainment, and telecom sector. Geopolitical events were considered a distant sixth by boards across all industries, but were ranked as the top risk by those in real estate, hospitality, and construction sectors—an assessment that would look significantly different in 2023 given the changed geopolitical context.

---

67   S. Sutherland, T. Dekker, and K. Pederson. 2021. The Board Imperative: Is Now the Time to Reframe Risk as Opportunity? *Ernst & Young.* 14 July. https://www.ey.com/en_gl/risk/is-now-the-time-to-reframe-risk-as-opportunity.

Remarkably, sustainability risks associated with climate change and natural resource constraints were rated as only the ninth most significant overall risk to their organization—with the notable exception of boards of energy and natural resources companies who ranked it first alongside changes in the regulatory environment. By August 2022, ESG risks, once a fringe activity, had become central to how organizations create and protect value for their stakeholders, elevating the role of the chief sustainability officer, and requiring robust processes for collecting and disclosing ESG data as part of overall risk management.[68]

## 3.2. The Risk Confidence Gap

As organizations accelerate investments in new digital products and services, decision-makers are demanding greater visibility into the connections between strategy, operations, and technology. To successfully mitigate increasingly complex risks, new risk management capabilities and investments are required to encourage a digital-first approach (Figure 13). The key question is whether existing technologies, methods, and skills can meet rising demands to manage this array of risks.

Drawing on recent surveys of board members, senior executives, and risk leaders, a majority of respondents expressed concern that the capabilities of most risk teams had not kept pace with rising demands, changes in business strategy, and emerging and atypical risks. Only a fifth of board members were confident that their organization had a highly effective disaster response and contingency plan, with just 13% of respondents expressing confidence that risk and compliance activities had effectively been embedded in the organization. Despite the growing necessity for risk teams to participate in major digital transformation projects, Accenture's 2021 Global Risk Management Study revealed that less than a third of respondents

**Figure 13: Developing an Integrated View of Digital Risk**

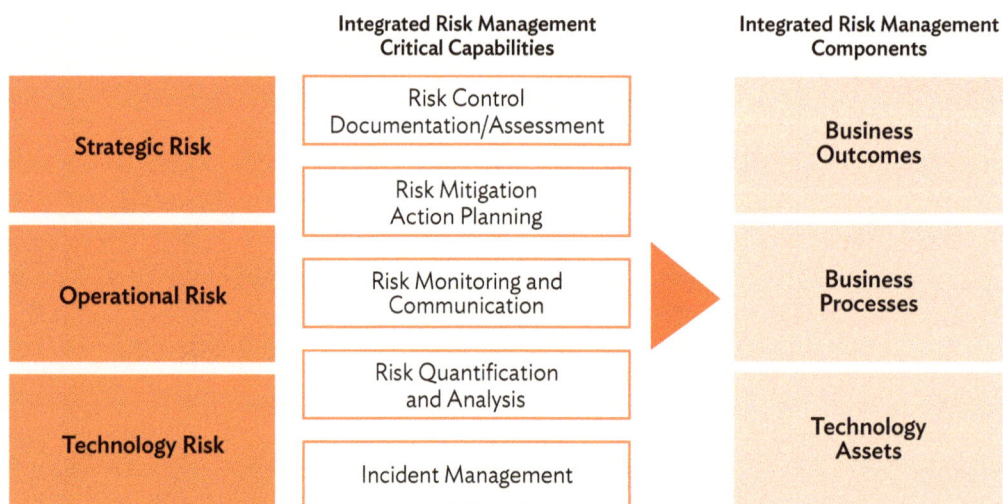

Source: J. Wheeler. 2020. 20 for 20: IRM Critical Capabilities and Top 20 Functions. *Gartner* (blog). 20 October.

---

68   S. Sutherland. 2022. The Board Imperative: Partner with CSOs to drive value-led sustainability. *Ernst & Young*. 9 August. https://www.ey.com/en_gl/board-matters/the-board-imperative-partner-with-csos-to-drive-value-led-sustainability.

were "very satisfied" with their ability in doing so. Only 49% of respondents said there were "fully capable" of assessing risks associated with cloud computing, with even smaller numbers reporting readiness for AI, blockchain, and other new technologies. [69]

This shortfall in risk management capabilities stands in contrast to rising risk demands—revealing an important gap in risk confidence and the potential cost of uncovered risk exposure.

A combination of factors is contributing to this confidence deficit:

- **Flawed risk management strategy.** Many board directors held the view that traditional governance, risk, and compliance tools do not provide an effective risk management strategy, fail to grasp the full scope and impact of risk throughout the organization, and leave organizations unprepared to address atypical and emerging risks (Table 4). Risk leaders are often not aligned with senior leadership about the upside potential of DX programs and differ in their views on strategic opportunities.[70] A Deloitte survey found nearly two-thirds of organizations are struggling to define their strategic risk appetite, which suggests that risk management strategies need to be revised at the highest level of the organization.[71]

- **Poor data infrastructure.** Organizations are also not well positioned to support their risk management efforts with timely and accurate risk data. Many risk teams are unable to leverage data and technology to deliver up-to-date, insight-driven reporting to senior leadership. Just half of chief executive officers surveyed were satisfied that their risk assessments—one of the most time-consuming and complex processes in risk management—were sufficiently data-driven (footnote 74) and offered insights into risk trends, threats, and mitigating strategies.[72]

**Table 4:** Legacy Risk Programs vs. Integrated Risk Management Models

| Solution Characteristics | Legacy Governance/Risk/ Compliance Methods | Integrated Risk Management |
| --- | --- | --- |
| Content | Compliance-driven | Risk-focused |
| Market definition | Ubiquitous | Targeted, purposeful |
| Design | Technical, control-based | Business-oriented, process-based |
| Features/functions | Rigid | Flexible |
| Use | Internally driven, departmental | Ecosystem-driven, cross-business unit, partners/suppliers |
| Architecture | Closed, proprietary | Open, integrated |
| Buyers/influencers | Technical practitioners | Business leaders |

Source: Gartner.

---

[69] S. Culp. 2021. In a world of risk, pace comes from preparation. *Accenture*. 13 July.

[70] Risk managers can serve as impartial advocates for the right balance of investment and the right pace of adoption. And they can explore the benefits of new technologies while helping identify the risks associated with these technologies. Quantum computing provides a good example of a technology with great promise but with new security risks, in this case related to quantum computing's ability to overwhelm current encryption protocols. S. Culp. 2022. Confronting the Risks of Innovation and Technology. *Forbes*. 27 September. https://www.forbes.com/sites/steveculp/2022/09/27/confronting-the-risks-of-innovation-and-technology/.

[71] J.H. Caldwell. 2021. A Moving Target: Refocusing Risk and Resiliency amidst Continued Uncertainty. *Deloitte Insights*. 1 February. https://www2.deloitte.com/us/en/insights/industry/financial-services/global-risk-management-survey-financial-services.html.

[72] *Ernst & Young*, "The Board Imperative," June 2021.

- **Lack of agility.** Risk velocity—the measurement of how fast an exposure can impact an organization—has become a third area of concern. The impact of the restrictions placed by the COVID-19 pandemic response underscored gaps among many organizations and an inability to respond effectively to the disruption. Most risk leaders acknowledge difficulties of maintaining reliable data to inform risk-based decisions. Ultimately, risk teams that are slow in uncovering new risks and managing risk workflows will become a liability for organizations that need to be able to respond to future crises in a timely and agile manner.

- **Lack of organizational integration.** Most organizations have risk oversight functions in areas like cybersecurity, information security, and regulatory compliance. Despite this, Deloitte reports that two-thirds of organizations do not have risk controls embedded within their business units. Too often assurance groups work in silos when the day-to-day reality requires working closely with one another to execute coordinated risk management activities. An AuditBoard poll of over 1,000 risk professionals found that nearly 60% of respondents have limited or no visibility into the issues identified by other risk groups, and nearly 70% lack a consistent reporting channel to senior management.[73] Taken together, these factors contribute to assurance fatigue and create dangerous gaps in risk coverage.

## 3.3. Managing Digital Risks: Toward and Integrated Approach

Digital risks are interconnected with other risk domains (e.g., strategic, operational, and technological), and usually cross over into other organizations (e.g., third-party service providers, business partners). Crucially, because risks often intersect—a single risk event can expose an organization to a variety of strategic, operational, and technological risks—the underlying causes are also frequently linked. Integrated risk management (IRM) aims to bridge the gaps between these interconnected risk domains focusing on compliance, resilience, performance, and risk assurance (Figure 11):[74]

(1) **Compliance.** While meeting compliance requirements is an increasingly complex endeavor, identifying and remediating areas of noncompliance has become a much more challenging task for many organizations. Risk teams need to be prepared for new laws and regulatory mandates; the rapid disclosure of noncompliance events; and stricter enforcements related to data privacy, labor relations, and ESG-related issues.[75] Without an integrated approach to risk management, mistakes will result in penalties and potential reputational damage.

---

[73]  J. Wheeler and A. Bhakta. 2022. The Business Resilience Gap: A Tipping Point. *AuditBoard*. 1 February. https://www.auditboard.com/blog/business-resilience-gap-tipping-point/.

[74]  IRM is a set of practices, processes, and enabling technologies that improves decision-making and performance through an integrated view of how well an organization manages its unique set of risks. Gartner. Gartner Information Technology Glossary: Definition of Integrated Risk Management. https://www.gartner.com/en/information-technology/glossary/integrated-risk-management-irm.; and Metricstream. Implementing Enterprise Risk Management Program: Step-by-Step Guide for Energy and Utilities Organizations. https://www.metricstream.com/learn/erm-enterprise-risk-management-energy-utilities.html.

[75]  KPMG. 2022. A Triple Threat across the Americas: 2022 KPMG Fraud Outlook. https://assets.kpmg.com/content/dam/kpmg/xx/pdf/2022/01/fraud-survey.pdf.

(2) **Resilience.** Resilience is focused on the ability of organizations to quickly identify, respond to, and recover from a digital risk event such as supply chain failure, cyber attack, system outage, etc. What is common across risk events is the need for awareness and understanding of what is most critical for preserving or restoring business continuity. IRM teams can conduct business impact analysis and carry out scenario exercises to identify plausible risks, identify areas requiring focused attention, and prepare possible responses.

(3) **Performance.** Managing and measuring the organization's overall performance by senior management can leverage IRM to ensure strategic alignment and operational performance of digital transformation efforts, covering third-party networks, sustainability, and risk disclosures.[76]

(4) **Risk assurance.** In terms of risk assurance, how does an organization know it is mitigating the right risks in the right way? An integrated approach offers the framework for defining risk appetite, establishing risk metrics, and monitoring risk on a continuous basis. Organizations that rely on outdated, legacy governance, risk, and compliance platforms cannot provide an appropriate level of assurance due to their siloed views of risk. What may be acceptable in one silo, such as IT security, may not support the intended level of risk mitigation in other areas, like data privacy or cross-border regulation of data flows,[77] and lead to a misallocation of resources to areas that may not be as critical. By having an integrated risk focus, organizations can more effectively analyze the total risk and prioritize their efforts accordingly.

IRM frameworks also map the respective functional responsibilities, which is indicative for the growing complexity of the digital risk landscape both within organizations as well as in relation to its external environment.

In summary, the spectrum of digital risks is growing more complex, challenging traditional, compliance-driven, internally focused approaches. Governments, business leaders, and risk practitioners can seize this opportunity to shift toward a more integrated approach and pursue collaboration beyond the organization with both public and private partners in the digital ecosystem.

## 3.4. Digital Risk Management Policies and Practices

Digital risk management requires a proactive effort to discover and mitigate digital risks, be it to protect critical infrastructure systems or to ensure that digital transformation initiatives can proceed. Effective risk protection involves anticipating all possible digital risks so that preventative steps can be taken to safeguard ongoing operations or the onboarding of a new technology.

Digital risk management differs from traditional business risk management in several respects. A first distinction concerns the rapid, location-independent diffusion of risk. Take the case of a ransomware attack that can shut down an entire municipality for days (or even weeks) in ways that a protest demonstration in

---

[76] Frameworks such as Integrated Reporting and the International Sustainability Standards Board disclosure requirements are set to become the norm for corporate reporting and will rely on IRM for better risk disclosures. Integrated Reporting. Integrated Reporting Framework. https://www.integratedreporting.org/resource/international-ir-framework/.; International Sustainability Standards Board. About the International Sustainability Standards Board. https://www.ifrs.org/groups/international-sustainability-standards-board/#about.

[77] Baker McKenzie. 2021/2022 Digital Transformation & Cloud Survey: A Wave of Change. https://www.bakermckenzie.com/-/media/files/insight/publications/2021/12/2021-digital-transformation--cloud-survey--a-wave-of-change.pdf?la=en.

front of a government office could not. A second characteristic involves the growing interconnectedness of internal and external business entities. A restaurant chain, for instance, can lose vast amounts of business if its third-party delivery partner changes its business model—as has happened in many recent cases. A third difference concerns specific risks about the handling, storage, and sharing of data. Another criterion that sets digital risk management apart is the expansive role of digital assets compared with the finite and relatively stable number of assets, employees, and business partners in traditional businesses. Digital business models, by contrast, need to manage a very large number of assets, such as IoT devices or autonomous systems driven by algorithms. The latter provide opportunities for automation which reduce human involvement, bringing about efficiencies and reducing human error but also come with new risks that require monitoring and adequate defenses. Finally, digital risk management needs to contend with the fact that digital business models can potentially reinforce socioeconomic biases at a massive scale and increase exclusion.

Governments can manage the digital risk of critical infrastructure through various initiatives and policies. The OECD Policy Toolkit on Governance of Critical Infrastructure Resilience offers a comprehensive policy framework to strengthen critical infrastructure resilience, ensure improved continuity of essential services, and overcome related governance challenges by adopting a systems approach based on partnerships between governments and critical infrastructure operators.[78] In the United States, for instance, the White House issued a National Security Memorandum on Improving Cybersecurity for Critical Infrastructure Control Systems in 2021.[79] The memorandum directs agencies to develop performance goals for critical infrastructure and establishes a voluntary, collaborative effort between the federal government and the critical infrastructure community to significantly improve the cybersecurity of these critical systems. The primary objective of this initiative is to encourage the deployment of technologies and systems that provide threat visibility, indications, detection, and warning, and that facilitate response capabilities in essential technology networks and control systems. Australia has implemented several initiatives to manage the digital risks of critical infrastructure. The critical infrastructure resilience strategy[80] brings together legislative and regulatory settings with industry and community partnerships through the trusted information-sharing network.

Managing the entire spectrum of digital risks at once is neither efficient nor effective. Not all risks have the same level of impact on desired outcomes. A preferred approach would be to focus on the most critical risk areas threatening the health of the organization. Uninterrupted operations, cyber attacks, third-party risks, and data leaks rank among the top risk areas, not only because their exploitation has the most negative repercussions, but also because they spill over into other digital risk domains. A pragmatic approach along these lines could be sequenced as follows:

1. Create a **digital footprint** by identifying all exposed technology assets within the organization and its broader digital ecosystem.
2. Develop an **incident response plan** to help focus and prepare the organization for cybersecurity incidents. This is most effective when paired with **crisis management** planning that pays attention to the interconnected nature of cybersecurity with other risk domains. Effective digital

---

78  OECD. 2019. Policy Toolkit on Governance of Critical Infrastructure Resilience. in *Good Governance for Critical Infrastructure Resilience*. Paris: OECD Publishing. https://doi.org/10.1787/fc4124df-en.

79  The White House. National Security Memorandum on Improving Cybersecurity for Critical Infrastructure Control Systems. https://www.whitehouse.gov/briefing-room/statements-releases/2021/07/28/national-security-memorandum-on-improving-cybersecurity-for-critical-infrastructure-control-systems/.

80  Cyber and Infrastructure Security Centre. Critical Infrastructure Resilience Strategy. https://www.cisc.gov.au/.

management does not prevent cyber threats but instead empowers organizations to maintain control when threats occur and informs relevant stakeholders about the risk response. Preparing for data breaches, for instance, can be achieved by setting out clearly the specific responses for each cyber threat scenario.

3. **Reduce the attack surface** to an absolute minimum. Compiling a detailed inventory of all technology vulnerabilities in the organization's digital ecosystem helps identify opportunities for reducing attack surfaces. A deeper analysis could involve risk assessments for both internal infrastructure and third-party networks which account for almost 60% of all data breaches. Other options, which will be discussed in Chapters 4 and 5, for reducing attack surfaces include the implementation of a zero trust architecture, multi-factor authentication, isolation of data backups from network access, or network segmentation with multiple firewalls.

4. **Monitor all network traffic** and create strict access policies for all sensitive technology assets. By following the Principle of Least Privilege, network access can be limited to those who absolutely require it.

5. With all exposed assets and their vulnerabilities mapped out, implement advanced **threat detection and response solutions** at the device and/or the network level.[81]

Effective risk management practices need to extend beyond basic crisis management to understand not only the organization's strategic risk landscape, but also how risks in executing this strategy match up against the organization's risk appetite and can impact overall results. Defining the organization's risk appetite is essential for effective risk management and needs to draw on the views of board members, executives, risk leaders, and operational teams.[82] Understanding the objective of a digital transformation initiative is a critical starting point for measuring progress and determining success or failure. To identify the risks that will prevent the organization from achieving its key strategic objectives, risk teams can rely, for instance, on the widely used COSO Integrated Framework.[83]

Risk management draws on measurable, quantifiable key risk indicators, which ideally tie one or more specific risks to a strategic objective. Technology platforms such as business intelligence dashboards can assist in visualizing the organization's risk tolerance levels through risk heat maps. Stress testing the organization's risk strategy is important to identify weaknesses and to improve the agility in reacting to risk events. Key performance indicators can monitor, for instance, the effectiveness of responding to a cybersecurity breach by quantifying time to detection and efficiency of response.

Risk functions alone cannot single-handedly manage risk. It is critical to integrate them with other assurance functions by sharing data and insights as well as engaging with operational teams. A unified assurance framework helps to define the roles of each assurance group (internal audit, risk management, compliance,

---

[81] Threat detection and response solutions include Endpoint Detection and Response which help detect and remediate threats at the endpoint or device level, Network Detection and Response that helps detect and remediate threats at the network activity level or communication between different devices, and Extended Detection and Response solutions that collect and analyze data from multiple level like email, endpoints, servers, cloud workloads, and networks.

[82] J. Kaplan et al. 2022. Creating a Technology and Cyber Risk Appetite Framework. *McKinsey & Company*.25 August. https://www.mckinsey.com/capabilities/risk-and-resilience/our-insights/cybersecurity/creating-a-technology-risk-and-cyber-risk-appetite-framework.

[83] C. Williams. 2019. COSO ERM Framework–Background & Overview. *Strategic Decision Solutions* (blog). 11 March. https://strategicdecisionsolutions.com/coso-erm-framework/.

etc.), clarify their working relationships, and agree on shared goals. These efforts prepare the stage for combining workflows, threat intelligence, knowledge, and technology platforms into a single enterprise-wide risk assessment and help to transition to a corporate reporting and tracking system.

To find out how organizations undertake digital risk management, AuditBoard conducted a Digital Risk Maturity Survey with over 125 risk managers in April 2022.[84]

A key finding was that, while most risk teams (90%) are increasingly aware of digital risks, most risk managers (63%) are still in the early stages of defining and assessing their digital risks. Only 30% of those surveyed are at a stage where they are actively mitigating digital risks. In this transition stage, they may misinterpret digital risk narrowly, when, in practice, digital risk is closely associated with business risks and strategic initiatives. In fact, most respondents have placed ownership of digital risks with functions outside of business operations (such as IT or security), which can lead to a misclassification of risks as technical or compliance-only and can give rise to fragmented risk management efforts.

Organizations can reach higher maturity levels in their digital risk management practices by transitioning from qualitative judgments to quantitative metrics, from manual to automated assessments, and from narrow technical and compliance-driven techniques to an integrated risk management framework.

An important step for reaching higher levels of digital risk maturity is to ensure that the practice is embedded in an organization-wide risk culture and closely aligned with operations. Organizations with mature risk practices apply an integrated risk management approach to assess, manage, and mitigate risk across all business units and risk and compliance functions, as well as key business partners, third-party suppliers, and malicious actors.

In a mature state, decision-makers rely on digital risk monitoring to collect and analyze data from an array of internal and external sources and draw on key risk indicators to alert management to areas requiring deeper scrutiny. Digital risk metrics need to reflect the desired outcomes for a given digital transformation initiative. For instance, is the intention to digitalize a single manual process, or does the digital transformation project involve the launch of a new business or service delivery model? According to AuditBoard's Digital Risk Maturity Survey, most organizations (84%) do not yet use reportable metrics to manage digital risk.

As organizations move up the maturity ladder, risk-relevant metrics should be introduced throughout all stages of digital transformation initiatives. Metrics at the decision-making stage can flag product or service viability issues. Implementation metrics provide valuable information to decision-makers looking to manage competing priorities. Monitoring provides a measure of success in reaching the target set by key stakeholders. Ultimately, these metrics provide insights into different aspects of digital risk, keeping in mind that not every digital risk is a "key risk" and not every metric is mission critical.

---

84    AuditBoard. Digital Risk Maturity Report 2022: Turning Digital Risk into Your Competitive Advantage. https://www.auditboard.com/resources/ebook/digital-risk-maturity-report-2022-turning-digital-risk-into-your-competitive-advantage/.

Digital technologies can help automate risk management. At the most advanced level of risk maturity, an automated IRM system provides risk monitoring, flags risks that require further inquiry, and triggers action steps for assurance teams. Digital risk management is by nature a collaborative effort that requires inputs from different teams and multiple systems. One of the most time-consuming activities for risk teams is to collect these inputs manually, using emails, surveys, and spreadsheets. Risk management technology solutions, enabled by software-as-a-service solutions, robotic process automation, and advanced analytics, can gather feedback from multiple stakeholders and systems more efficiently and promote more effective risk management practices.

**Table 5:** Key Capabilities for Integrated Digital Risk Management

| Risk and Control Documentation/Assessment | Risk Mitigation Action Planning | Risk Quantification and Analytics |
|---|---|---|
| 1. Risk-related content, including a risk framework, taxonomy/library, KRIs catalog, and legal, regulatory, and organizational compliance requirements<br>2. Risk assessment methodology and calculation capabilities (e.g., quantitative risk ratings)<br>3. Policy documentation and control mapping<br>4. Documentation of workflows, including authoring, versioning, and approval<br>5. Business impact analysis<br>6. Audit work and testing management<br>7. Third-party control evaluation | 8. Project management capabilities to track progress on risk-related initiatives, audits, or investigations<br>9. Risk control testing capabilities, such as continuous control monitoring | 13. Machine learning or other AI-enabled analytics<br>14. "What if" risk scenario analysis capabilities<br>15. Statistical modeling and predictive capabilities (e.g., Monte Carlo simulation, value at risk, statistical inference) |
| | **Risk Monitoring and Communication** | **Incident Management** |
| | 10. Risk scorecard/dashboard capabilities<br>11. External data integration (e.g., information security vulnerability, assessment data)<br>12. Ability to link Key Risk Indicators KRIs to performance metric | 16. Incident data capture<br>17. Incident management workflow and reporting<br>18. Root cause analysis<br>19. Crisis management<br>20. Investigative case management |

KRI = key risk indicator.
Source: J. Wheeler. 2020. 20 for 20: IRM Critical Capabilities and Top 20 Functions. *Gartner* (blog). 20 October. https://blogs.gartner.com/john-wheeler/top-20-irm-functions-features/.

## 3.5. The Upside of Effective Digital Risk Management

The upside of higher investments in digital risk management is that organizations that discover these new digital risks early and can address them effectively are likely to gain advantages through better performance and stronger resilience.

Another consideration for investing in risk management technologies is to keep pace with the evolving digital risk landscape. Cloud-based risk management tools have become the software of choice due to their ability to consume large amounts of data and enable other forms of system integration. Integrating cloud-based technology with other systems is especially relevant when tracking third-party digital risk, which can be collected from external data sources (e.g., security rating services, regulatory disclosures, social media), interviews, questionnaires, independent reports, and targeted audits.

Organizations can turn digital risk management into their competitive advantage. An organization that successfully selects and implements frontier technology solutions to support its integrated risk practice can aim for efficiency and effectiveness gains in its digital risk management program. In turn, a strong digital risk program can enable an organization to scale its digital transformation initiatives thanks to increased confidence levels among executives.

Decision-makers need better visibility and understanding of digital risks, including the growing ecosystem of third-party providers. Applying an integrated risk management approach can reduce silos and provide a more complete and objective view across strategic, operational, and technological risks. This information can also be used to shape plans by taking advantage of scenario analysis and predictive modeling. While cybersecurity and regulatory change represent risks that may lead to losses, digital risk is directly linked to the organization's strategic mission and future growth. Attaining this level of digital risk management is a mission-critical proposition and a core function for digital risk teams.

**Equipping economies.**
Importance of cybersecurity awareness to navigate the digital landscape securely (photo by ADB).

# 4

## CYBERSECURITY RISKS

Cybercrimes topped the global risk barometer in 2023[85] and rank as a top-five risk in East Asia and the Pacific as well as in Europe.[86] Cybercrime is costing public institutions and private enterprises hundreds of millions of dollars in ransom payments and damages every year, and is putting critical infrastructure, societal cohesion, and mental health at risk.

Cybersecurity failure is one of the risks that worsened the most during the COVID-19 pandemic. Ransomware attacks rose by 435% in 2020 compared to 2019.[87] Criminals with little technical knowledge have been able to subscribe to ransomware software for a low monthly subscription and use cryptocurrency to evade detection. There were around 236 million ransomware attacks globally in the first half of 2022. Statista estimates that about 70% of businesses suffered one or more ransomware attacks in 2022, with education, government, and health care as the top three sectors being targeted. These attacks directly affected workflow and employees. Almost 40% of enterprises had to lay off employees after an attack, and 35% experienced C-level resignations. Another 33% had to suspend operations temporarily. Most organizations (63%) also reported that the attackers were in their network for up to 6 months before detection, while 21% said attackers had 7–12 months of unauthorized access, and 16% said the attackers were in the network for over a year. Almost half of the organizations (49%) paid ransom to prevent revenue losses, and another 41% paid ransom to quicken the recovery process. Of those with reported losses, most (67%) said they lost between $1 million and $10 million, and over half of the companies that paid the ransom had corrupted data after removing encryption.

The escalation of cyber threats is outpacing societies' ability to effectively prevent and manage them. Research found that two-thirds of cyber attacks are coming from vulnerabilities that software makers had previously detected but failed to patch. At the same time, the demand for capabilities based on converging technologies—including AI, IoT-enabled devices, edge computing, blockchain and 5G—is only growing. Technology platforms, tools, and interfaces connected via an increasingly decentralized and fragmented internet are creating a more complex cyber-threat landscape and a proliferation of failure points.[88] Active IoT connections, ranging from smart home appliances to agricultural sensors, are expected to increase to 14.4 billion by end-2022.[89] In response to the lack of cyber protection of IoT products, draft EU legislation will require IoT makers to obtain mandatory certificates that show they are meeting basic cyber safety requirements.[90]

This chapter will provide a macro perspective on cybercrime, clarify key concepts, and outline a number of frameworks that are being used to assess the maturity of countries and organizations to prepare for and respond to cyber threats. Until recently, the niche nature of cybersecurity on the global policy agenda stalled

[85]  Allianz. Allianz Risk Barometer. https://www.agcs.allianz.com/news-and-insights/reports/allianz-risk-barometer.html.
[86]  World Economic Forum. Global Risks Report 2022. https://www.weforum.org/reports/global-risks-report-2022/.
[87]  *GRC World Forums*. 2021. Ransomware Demands Soar by 518% in 2021. 13 August. https://www.grcworldforums.com/ransomware/ransomware-demands-soar-by-518-in-2021/2357.article.
[88]  A much-hyped technology innovation is the metaverse, which promises unprecedented interoperability and immersive virtual reality experience. Users will be required to navigate security vulnerabilities inherent in these sophisticated technologies that often characterize decentralized blockchains and lack of structured guardrails or sophisticated onboarding infrastructure.
[89]  S. Sinja. 2023. State of IoT 2022: Number of Connected IoT Devices Growing 18% to 14.4 Billion Globally. *IoT Analytics* (blog). 18 May. https://iot-analytics.com/number-connected-iot-devices/.
[90]  J. Espinoza. 2022. EU to Impose Tough Rules on "Internet of Things" Product Makers. *Financial Times*. 7 September.

the integration of cyber capacity building into the development agenda. The second part of the chapter puts a spotlight on how to achieve a closer partnership between cybersecurity communities and digital development practitioners.

# 4.1. Macro-Perspectives

Global cybercrime is a $6 trillion industry, [91] which would make it the world's third-largest economy after the US and the PRC.[92] With cyber attacks on the rise, annual costs related to cybercrime are projected to increase by 15% to reach $10.5 trillion a year in 2025.[93]

The scope of the threat is growing. The proliferation of cyber-based conflicts (Table 6) between nation-states and the growing fusion between cyber and physical or kinetic confrontations suggest that the world may be inching closer to "advanced cyberconflict."[94] Nations are prepared to devote significant resources toward achieving strategic advantages in cyberspace. With dedicated research programs aimed at developing new kinds of cyber attacks, the stockpiling of "exploits," and the combination of attack tools and techniques, the complexity and sophistication of methods used by state actors and private cyber criminals have increased.

The large-scale penetration of US cybersecurity in 2020 by way of the SolarWinds hack and the shutdown of the US Colonial Pipeline along the East Coast Corridor system in response to a ransomware attack are just two of the most notorious recent examples at the time of publication.[95] The line between state and non-state actors continues to blur, with banking, energy, defense, and health care among the leading targets for attacks by state actors. The Carnegie Endowment for International Peace and the Threat Intelligence unit of BAE Systems have documented a timeline, which chronicles ~200 cyber incidents targeting financial institutions since 2007.[96] Cybersecurity attacks on financial institutions have become more frequent, sophisticated, and destructive in recent years and, according to the G20 in 2017, could "undermine the security and confidence and endanger financial stability."

---

[91]  L. Brown. 2021. The New Threat Economy: A Guide to Cybercrime's Transformation–and How to Respond. *Security Magazine*. 15 June. https://www.securitymagazine.com/articles/95387-the-new-threat-economy-a-guide-to-cybercrimes-transformation-and-how-to-respond.

[92]  Typical sources of revenue include trade secret theft and data trading, theft in currency, digital money laundering, and the lucrative (albeit legal) industry of building cybersecurity tools.

[93]  J. Boehm et al. 2022. Cybersecurity Trends: Looking over the Horizon. *McKinsey*. 10 March. https://www.mckinsey.com/business-functions/risk-and-resilience/our-insights/cybersecurity-trends-looking-over-the-horizon#.

[94]  Advanced cyberconflict portrays a situation where nations begin to engage in repeated digital attacks and counterattacks, blurring the line between online and offline targets and increasingly targeting physical assets, like power grids or water supplies.

[95]  *This Is How They Tell Me the World Ends*. https://thisishowtheytellmetheworldends.com/ In this Financial Times Book of the Year 2021, New York Times Journalist Nicole Perlroth describes the evolution of the global cyber weapons market and offers an inside look at global cyberattacks and its actors.

[96]  Carnegie Endowment for International Peace. Timeline of Cyber Incidents Involving Financial Institutions. https://carnegieendowment.org/specialprojects/protectingfinancialstability/timeline (accessed 2 November 2022). The database is searchable by country, year, attribution, incident type, and estimated financial damage.

**Table 6: The Upward Trajectory of Cyber Attacks**

| Escalation | |
|---|---|
| • **100% rise** in significant nation-state incidents between 2017 and 2020 <br> • **10 publicly attributed cyber attacks/month** in 2020 | • **Over 40%** cyber attacks had physical and digital components <br> • **20%** of cyber attacks are correlated with regional conflicts |

**Intensification**

- **Most common weapons:** Surveillance (50%); Network incursion and positioning (15%); Damage or destruction (14%); Data extraction (8%)
- **Supply chain attacks:** 78% increase (2019); 27 known nation state sponsored supply chain attacks between 2017–2020; 40% of supply chain security breaches are now indirect;
- **Known Nation-State targets:** Enterprises (35%); Cyber defense (25%); Media and communications (14%); Government bodies and regulatory agencies (12%); Critical infrastructure networks (10%)

**Expansion**

| | |
|---|---|
| • 20% of cyber attacks involved sophisticated weapons <br> • 50% of cyber attacks involved low-budget tools <br> • 58% of experts state that the recruitment of cyber criminals by nation-states has become common practice. | • 45% rise in cyber attacks on health-care organizations since November 2020 <br> • 50% rise in cyber attacks on pharmaceutical companies during COVID-19 <br> • 50,000+ new COVID-19-related websites created between February and May 2020 involved fraud |

**Cyber War or Cyber Peace?**

- 70% of experts believe a cyber treaty is necessary to prevent cyber escalations
- … however, only 15% are confident that an agreement will be reached in the next 5–10 years
- 30% believe there is no prospect of any cyber treaty within any timeframe

Source: M. McGuire. 2021. Nation States, Cyberconflict, and the Web of Profit. *HP Wolf Security* (blog). 8 April.

According to recent threat assessments by the US Government's Director of National Intelligence, nation-states use cyber operations to "steal information, influence populations, and damage industry, including physical and digital critical infrastructure."[97] Many major transnational cybercrime groups engage in direct wire-transfer fraud from victims or use other forms of extortion alongside ransomware. Business e-mail compromise, identity theft, spoofing, and other extortion schemes rank among the top-five most costly cybercriminal schemes (footnote 98).

A web of profit connects underground cybercrime economies across the world and is shaping the character of nation-state conflict within online environments. Not only are many nation-states making active use of available tools and techniques, but some are also recruiting cybercriminals to act as proxies to further their interests. Conversely, many tools originating from national security agencies are finding their way into the hands of cybercriminals. The US National Security Agency's Eternal Blue exploit, which was used by the WannaCry hackers to cause disruption worldwide in 2017, serves as an infamous example.[98]

---

[97] Office of the Director of National Intelligence. 2021 Annual Threat Assessment of the US Intelligence Community. https://www.dni.gov/index.php/newsroom/reports-publications/reports-publications-2021/item/2204-2021-annual-threat-assessment-of-the-u-s-intelligence-community.

[98] L. Hay Newman. 2018. The Leaked NSA Spy Tool That Hacked the World. *Wired*. 7 March. https://www.wired.com/story/eternalblue-leaked-nsa-spy-tool-hacked-world/.

The COVID-19 pandemic response over the last 2 years has exacerbated cyber threats, whether in seeking to steal intellectual property on vaccines or targeted attempts to disrupt supply chains. Attackers will continue to perfect the use of AI, machine learning, and other technologies to launch increasingly sophisticated attacks, which will expedite—from weeks to days to hours—the end-to-end attack life cycle.

No organization is immune. Federal and state governments, municipalities, academic institutions, and SMEs face such risks along with large companies. On-demand access to data and information platforms is growing. In 2020, on average, every person on Earth created 1.7 megabytes of data each second. With the growing importance of the cloud, the firms providing cloud services—first and foremost large hyperscalers that operate globally—are responsible for storing, managing, and protecting an increasing share of this data.

To manage the exponentially growing data volumes, companies are implementing new technology platforms, including data lakes that aggregate information across environments. Recent high-profile attacks exploited this expanded data access. Examples are the Sunburst hack in 2020 that spread via malicious code to customers during regular software updates,[99] or the use of compromised employee credentials from a third-party application to access more than 5 million Marriott hotel guest records. This attack followed on the heels of one of the largest breaches in history,[100] which compromised the reservation records of 500 million people in 2018.[101]

Companies and government agencies are forced to make decisions in an uncertain environment to protect against a range of threats. The estimated average cost for organizations that fall victim to a successful attack is estimated at $1.85 million per incident. While the short-term focus is on the commercial impact of unauthorized access to data, in the longer term, organizations are more concerned with the damage to trust and reputation. According to one US-based survey, nearly 90% of respondents believe they have been cyber attacked by an organization acting on behalf of a nation-state, but only 27% have confidence in their ability to differentiate between nation-state cyber attacks and other cyber attacks.[102] The increase in cyber attacks has prompted calls to governments to step up support to organizations and protect critical infrastructure against cyber attacks.

This fusion of politics, strategic pursuits, commerce, and crime is raising new challenges for digital regulation. An increasingly complex and continuously evolving regulatory landscape and persistent gaps in resources, knowledge, and talent will continue to be outpaced by cybersecurity requirements.

[99] P. Shakarian. 2021. The Sunburst Hack Was Massive and Devastating–5 Observations from a Cybersecurity Expert. *GovTech*. 4 January. https://www.govtech.com/security/The-Sunburst-Hack-Was-Massive-and-Devastating--5-Observations-from-a-Cybersecurity-Expert.html.

[100] B. Barrett. 2020. Hack Brief: Marriott Got Hacked. Yes, Again. *Wired*. 31 March. https://www.wired.com/story/marriott-hacked-yes-again-2020/.

[101] Marriott's customer satisfaction scores dipped, suggesting that the breach may cause long-term harm to guest loyalty. Studies show that nearly a quarter of Americans will stop doing business with a company that has been hacked, while more than two in three people trust a company less after a data breach. J. Hollander. 2020. Marriott Data Breach FAQ: What Really Happened? Hotel Tech Report. 9 December. https://hoteltechreport.com/news/marriott-data-breach.

[102] The Center for Strategic and International Studies. 2022. Organizations and Nation-State Cyber Threats in the Crosshairs. https://www.csis.org/events/report-launch-organizations-and-nation-state-cyber-threats-crosshairs.

Cyber risk management has not kept pace with digital transformation and the accelerated data collection, sharing, sale, integration, and advanced analytics. Many companies are at a loss on how to identify and manage digital risks—yet are facing stiffer compliance requirements due to growing privacy concerns, high-profile breaches, and complex cross-border data flow regulations. Compounding the challenge, regulators are increasing their requirements for corporate cybersecurity capabilities—often with the same level of oversight that is being applied to credit and liquidity risks in financial services or to operational and physical safety standards for critical infrastructure. The lack of consensus on acceptable standards for online conduct or effective regulation underscores the argument that the internet poses far greater risks than generally understood leaving a void in the global governance architecture.

## 4.2. Key Cyber Security Concepts

**Cyber risks** hold the potential for financial loss, disruption, or damage to the reputation of an individual, organization, or government from failure, unauthorized or erroneous use, or other malicious exploitation of its information systems.[103] One way for these cyber risks to materialize is through cyber attacks carried out for a variety of reasons by different actors, who take advantage of digital vulnerabilities (see Table 7; Cybersecurity Glossary in Appendix 1)

**Cyber vulnerabilities** are specific weaknesses in a computer system, online network, or processes and procedures, such as a lack of encryption or poorly designed firewalls that allow an intruder to execute commands, gain unauthorized access to data, and/or conduct denial-of-service attacks (Table 8). Cyber vulnerabilities can also involve a deficit in capacity or skills protecting those systems or networks, for instance due to poor cyber hygiene.

**Cybersecurity** involves the management of cyber risks and vulnerabilities. Relying on people, systems, and technologies, the goal is to protect information stored in digital formats from being taken, damaged, modified, or exploited.

**Digital Information** can be acquired in a variety of ways, ranging from illicit cyber attacks on computer systems to the aggregation, analysis, and sale of social media information about individuals' online behavior through data brokers which could be used in political campaigns or to achieve social and political outcomes through disinformation—much of which can be achieved without doing anything explicitly illegal.

**Effective cybersecurity** requires adequate policies and strategies along with institutions that have the human and material resources to follow through on implementation. Governments, civil society, media outlets, and the private sector are well advised to consider cybersecurity in all operational aspects, including enterprise architecture, procurement, supply chains, and contracting agreements (footnote 132).

---

[103]  USAID. *Cybersecurity Primer.* https://www.usaid.gov/digital-development/usaid-cybersecurity-primer.

## Table 7: Cyber Attack Objectives

| Objective | Purpose | Examples |
|---|---|---|
| Acquisition | Intelligence | Acquisition of military, industrial, or political secrets and intelligence. Around 95% of cyber attacks in the manufacturing sector are now associated with acquisitional espionage. |
| | Data | Data breaches involving nation-state actors amount to approximately 40%, up from 20%–25% in 2018–2019. |
| | Revenues | Cyber attacks represent a lucrative source of revenue for cybercriminals and some nation-states. |
| Incapacitation | Sabotage | Damage or incapacitation of infrastructure assets (e.g., the Shamoon cyber attacks on Gulf oil companies using the Disttrack malware, which wiped files and rendered systems inoperative.) |
| | Disruption | Impairing network functionality, internet shutdowns, or network disruptions have been estimated to cost the global economy around $1 trillion (~1 % of global GDP) in 2020—50% more than predicted in 2018. |
| Shaping | Opinion | "Cognitive hacking," such as disseminating disinformation to spread social divisions or to disrupt electoral processes. |
| | Regime Change | Cyber interventions by Russian hackers against Ukraine's government, IT infrastructure. |
| Hybridization | Tactical Support | Use of cyber capacity to augment conventional means of attacking and adversary. |

GDP = gross domestic product, IT = information technology.
[a]  Microsoft. Microsoft Digital Defense Report 2022. https://www.microsoft.com/en-us/security/business/microsoft-digital-defense-report-2022.
[b]  K. Kizzee. 2023. Cyber Attack Statistics to Know in 2023. Parachute. 13 September. https://parachute.cloud/2022-cyber-attack-statistics-data-and-trends/.
Source: The Center for Strategic and International Studies keeps a timeline dating back to 2006 listing cyber attacks on government agencies, defense and high-tech companies, or economic crimes with losses of $1+ million. The Center for Strategic and International Studies. Significant Cyber Incidents. https://www.csis.org/programs/strategic-technologies-program/significant-cyber-incidents; M. McGuire. 2021. Nation States, Cyberconflict, and the Web of Profit. *HP Wolf Security* (blog). 8 April.

## Table 8: Illustration of Selected Cyber Weapons

| Objective | Cyberweapon | Complexity | Application |
|---|---|---|---|
| **Acquisition** | | | |
| Intelligence | Hardware backdoors | 4 | "Lojax" malware acts like a rootkit but attacks UEFI—the fundamental key to any computer. It can even survive reinstallations of operating systems and has been associated with cyber espionage operations. |
| Intelligence | RAT (Remote Access Trojan) | 3 | The APT10 group used the PlugX Rat in 2019 for cyber attacks against a Southeast Asian government and private sector organizations. |
| Data | Keyloggers | 3 | QUERTY keylogging malware is a plug-in for the NSA REGIN cyber weapon with mass surveillance applications. |
| Revenues | Ransomware | 2 | SamSam ransomware was linked to nation-state cyber attacks on cities (Atlanta, San Diego $6m+ extortion; >$30m damage); insertion of malicious code by DarkSide shuts US Colonial Pipeline down (75 Bitcoins, ~ $5m ransom). Cryptocurrencies are common means to make transfer ransom payments. |

*continued on next page*

**Table 8** *continued*

| Objective | Cyberweapon | Complexity | Application |
|---|---|---|---|
| Status | Logic Bombs | 2 | Malware planted on Sony Pictures. > 4,000 computers wiped clean following release of the satirical movie about the leader of the Democratic People's Republic of Korea. |
| **Incapacitation** | | | |
| Disruption | dDos/Botnet | 2 | 2018 dDoS attack on Github—one of the largest ever. Hit by 1.35 terabytes per second of traffic, causing temporary loss of service. Attributed to nation-state. |
| Disruption | Wiper Malware | 3 | The NotPetya malware, which emerged in late 2016 and spread rapidly in 2017 disguised itself as ransomware. In fact, it was an endpoint wiper aimed at causing maximum disruption to systems. |
| Sabotage | Worms and targeted malware | 5 | Disttrack worm: sophisticated malware used within the "Shamoon" cyber attacks to target Gulf oil companies, deleting data and disrupting operations |
| Disruption | Malware framework (multiple tools) | 4 | Triton malware was used to take over safety systems in Saudi petrochemical plants and Schneider Electric's safety system, which is used to initiate safe shutdown procedures in case of emergency. Attributed to one or more nation-states. |

Note: Level of complexity/sophistication: 1 (low) – 5 (high).
Source: M. McGuire. 2021. Nation States, Cyberconflict, and the Web of Profit. *HP Wolf Security* (blog). 8 April.

As the scope, volume, and sophistication of cyber attacks and digital disruptions have grown exponentially, cybersecurity preparedness, digital resilience, and cyber capacity building have become top priorities for both developed and developing countries. Israel, for instance, has established a comprehensive cyber capacity-building program which includes (i) the formation of specialized government agencies to monitor, mitigate, and respond to national-level cyber threats through national or sector computer emergency response teams, computer security incidence response teams, cyber security operation centers, and other expert centers; (ii) training regulators and government agencies to guide, control, regulate, and support their organizations and those they oversee; (iii) collaboration with other countries to share information and promote collective defense; and (iv) development of products, services, and capabilities to prevent, manage, and respond to cyber risks and to export Israeli know-how, innovation, technologies, and expertise abroad.[104] It should be noted that Israeli cyber-intelligence companies have been responsible for the development and export of some of the most notorious cyber-espionage tools, such as Pegasus, that was used for the surveillance of government critics, journalists, and political leaders from several nations around the world.[105]

[104] Israel National Cyber Directorate. Israel International Cyber Strategy. https://www.gov.il/en/departments/news/international_strategy; and W. van Pruisenweg. 2021. *Integrating Cyber Capacity into the Digital Development Agenda*. The Hague: Global Forum on Cyber Expertise.

[105] R. Bergman and M. Mazzetti. 2022. The Battle for the World's Most Powerful Cyberweapon. *The New York Times*. 28 January. https://www.nytimes.com/2022/01/28/magazine/nso-group-israel-spyware.html.

# 4.3. Cybersecurity Frameworks and Maturity Models

The global community has stepped up its efforts to diagnose gaps in the cybersecurity postures of countries, governments, and enterprises to make better decisions on interventions and investments to enhance cyber capacity. Research institutions, regional organizations, and companies have also developed frameworks, models, and tool kits and applied them across the globe.

Cybersecurity risk assessments are increasingly replacing checkbox compliance routines to become the foundation of a risk management strategy. Such assessments contribute to a better understanding of where countries stand in terms of cyber maturity and their preparedness to counter increasing cyber threats to governments, industry, and citizens. Risk assessments serve both to establish a baseline as well as to track risk mitigation. Cybersecurity maturity models (CMMs) can provide a road map for countries and organizations for measuring, assessing, and enhancing cybersecurity by learning from best practice, engaging with experts from diverse backgrounds, and taking a life-cycle and continuous improvement approach.

Most models use a five-level framework to assess the status of different domains, applying a hybrid maturity model: (i) physical security, (ii) people security, (iii) data security, (iv) infrastructure security, and (v) crisis management. Organizations at Level 1 have no security management in place and perform cyber security practices only in an ad hoc manner. Level 2 requires an organization to establish guiding policy objectives and document practices within a domain. By documenting practices, an organization can execute the practices in a repeatable manner to achieve intended outcomes. Level 3 requires an organization to establish, maintain, and resource a plan for managing cyber security activities. The plan may include information on missions, goals, project plans, resourcing, required training, and involvement of relevant stakeholders. Level 4 requires an organization to review and measure practices for effectiveness. Organizations at this level must also take corrective actions when necessary and inform senior management of status or issues on a recurring basis. To reach Level 5 maturity, an organization must have an automatic security management system in place, which standardizes and optimizes implementation throughout. The organization must develop procedures from standard guidance typically provided by senior management.

To implement best cyber security practices, standards, such as the National Institute of Standards and Technology (NIST) Cybersecurity Framework or the International Organization for Standardization (ISO) 27000 series, are applied to perform and measure security levels in all CMMs. Models differ by having different goals and security requirements to fill gaps between cyber risks, implementation, technical elements, and community involvement.

Leading cybersecurity maturity models developed by the Global Cyber Security Capacity Center, NIST, ISO/ International Electrotechnical Commission (IEC), the World Bank, the Department of Energy, and the Council for Registered Ethical Security Testers (CREST) are summarized below.

- The Global Cyber Security Capacity Centre at Oxford University is a research center focused on global cybersecurity capacity-building initiatives. Its Cybersecurity Maturity Model for Nations[106] defines a country's cybersecurity maturity from multiple perspectives to span (i) cybersecurity policy and strategy, (ii) cybersecurity culture and society, (iii) building knowledge and capabilities, (iv) legal and regulatory frameworks, and (v) cybertechnologies and standards. The CMM covers around 600 indicators to rate maturity on these five dimensions. Each dimension involves five stages of maturity, with the lowest level implying a non-existent, or inadequate, level of capacity, and the highest indicating both a strategic approach, and the ability to dynamically respond to environmental considerations, including operational, sociotechnical, and political threats.

  Deployment of the CMM is a multistep and multi-stakeholder process which includes desktop research; in-country focus group discussions with key stakeholders; and an in-depth, peer-reviewed report which describes the in-country cybersecurity context, summarizes the findings for each dimension, outlines the stages of cybersecurity capacity maturity, and provides recommendations. Since 2015, more than 120 CMM reviews in over 85 countries have been completed by the Centre and its implementation partners (including the Oceania Cyber Security Centre, ITU, the Organization of American States, GIZ, and World Bank), allowing comparisons of cybersecurity capacity across different countries and over time.[107] All CMM reviews, including links to country reports that have been published and case studies are listed on the Global Cyber Security Capacity Centre's website (e.g., Bangladesh, Brazil, Lessons Learned, Cybersecurity in Pacific Island Nations).

- NIST outlined its guidelines for conducting a cyber risk assessment in their Special Publication 800–30 (NIST SP 800–30).[108] The guidance outlined in NIST Cybersecurity Framework[109] has been widely applied across industries and companies of all sizes.[110] The value of using NIST SP 800–30 is that it gives risk management teams the ability to examine risks through the lenses understood by government and business leaders, detailing threat type, business impact, and financial impact. Expressing risks in this format helps bridge the gap between cybersecurity professionals and decision-makers.

  NIST has developed a robust ecosystem of guidance, tool kits, and supporting documentation to guide organizations regulated by the US federal government, which are widely used in many countries and have been translated into a dozen languages. NIST SP 800–30 uses a hierarchical model to indicate the extent to which the results of a risk assessment inform the organization, with each tier expanding to include more stakeholders across the organization.

---

[106] Global Cybersecurity Capacity Centre. Cybersecurity Capacity Maturity Model for Nations 2021 Edition. https://gcscc.ox.ac.uk/cmm-2021-edition.

[107] A 2020 evaluation found that CMM reviews increased cybersecurity capacity building, contributed to greater collaboration within government and with business and wider society, helped define roles and responsibilities, and increased funding.

[108] National Institute of Standards and Technology (NIST). NIST Special Publication 800-30: Guide for Conducting Risk Assessments. https://www.nist.gov/privacy-framework/nist-sp-800-30.

[109] For an overview of implementing the NIST Cybersecurity Framework see Reciprocity. Complete Guide to NIST: Cybersecurity Framework 800-53, 800-171. https://reciprocity.com/resource-center/complete-guide-to-nist-cybersecurity-framework-800-53-800-171/.

[110] NIST. Cybersecurity Framework Documents. https://www.nist.gov/cyberframework/framework.

In August 2023, NIST released a draft Cybersecurity Framework 2.0 for public comment. The proposed changes, which aim to improve cybersecurity practices, include adding new categories of security and privacy risk, integrating supply chain risk management, and making the framework easier to use for organizations of all types. Perhaps the most significant proposed change is the introduction of a "govern" function, which emphasizes that cybersecurity governance is critical to managing and reducing cybersecurity risk.[111]

■ The ISO and the IEC have jointly published the 27000 series documentation for risk management, specifically ISO/IEC 27005,[112] which supports organizations seeking to build a risk-based cybersecurity program. The ISO/IEC standard is more popular among organizations in the EU, but is applicable to organizations of all shapes and sizes. The series is deliberately broad in scope, covering more than just privacy; confidentiality; and IT, technical, and cybersecurity issues. The recommendations are broken down into the following areas: security risk assessment; security policy; asset management; human resources security; physical and environmental security; communications and operations management; access control; information systems acquisition, development, and maintenance; information security incident management; and business continuity management. Similar to NIST SP 800–30, using the ISO/IEC guidance is most beneficial for organizations pursuing or already maintaining an ISO/IEC certification.[113]

■ The Cybersecurity Capability Maturity Model is a US Department of Energy program to help critical infrastructure organizations evaluate and potentially improve their cyber security practices. This approach has been used to create cyber security capability models in the electricity, oil, and natural gas sectors. The model's architecture uses 10 security domains, with each domain featuring a structured set of cyber security practices. The model can be used to (i) strengthen organizations' cybersecurity capabilities, (ii) enable organizations to effectively and consistently evaluate and benchmark cybersecurity capabilities, (iii) share knowledge and best practices across organizations, and (iv) enable organizations to prioritize actions and investments to improve cybersecurity. The model is publicly available for download.[114]

■ The World Bank's Combating Cybercrime: Capacity-Building Assessment Tool 2017[115] was developed in partnership with seven other organizations, including the Council of Europe, the Republic of Korea, the International Telecommunications Union (ITU), the UN Conference on Trade and Development, and the UN Office on Drugs and Crime, to build the capacity of policymakers, public prosecutors, investigators, and civil society in the policy, legal, and criminal justice aspects to combat cybercrime. The tool can be used both as a stand-alone activity by countries to evaluate their current capacity (or lack thereof) to combat cybercrime and identify capacity building priorities as well as a due diligence assessment to enable operational teams to appraise a country's readiness to tackle cybercrime.

---

[111] NIST. Cybersecurity Framework 2.0. https://www.nist.gov/cyberframework.

[112] ISO2700 Information Security. ISO/IEC 27005 2022: Information Security, Cybersecurity, and Privacy Protection– Guidance on managing information security risks. https://www.iso27001security.com/html/27005.html.

[113] *Secureframe*. 2022. ISO 27000 series: What the Standards are + their Purpose. Blog. 17 March. https://secureframe.com/blog/iso-27000.

[114] Government of the US, Department of Energy. Cybersecurity Capability Maturity Model (C2M2). https://www.energy.gov/ceser/cybersecurity-capability-maturity-model-c2m2.

[115] An updated version of the assessment tool is scheduled to be completed by end–2022.

A unique aspect is that it serves as a self-diagnosis tool across nine dimensions: (i) Non-legal dimensions, such as national strategies, policies, and cooperation with the private sector; (ii) Legal framework; (iii) Substantive law, addressing criminal activities; (iv) Procedural law, mainly addressing investigatory matters; (v) e-Evidence; (vi) Jurisdiction; (vii) Safeguards, focusing on due process, data protection, and freedom of expression; (viii) International cooperation; and (ix) Capacity building at both institutional and staffing levels. The Microsoft Excel-based assessment tool, which includes 115 indicators, is available as a global public good.[116]

- The CREST[117] is an international not-for-profit accreditation and certification body that represents and supports the technical information security market. The CREST Cybersecurity Incident Response Maturity Assessment comprises a set of freely accessible, scalable spreadsheet tools that include (i) cyber threat intelligence assessment, (ii) cyber security incident response maturity assessment that assesses an organization's response capability, and (iii) penetration testing maturity tools.[118]

- The OECD Policy Framework on Digital Security of 2022 looked at the economic and social dimension of cyber security.[119] It develops and approach to digital security policy and is meant to equip policymakers to use the organization's digital security recommendations.

- The UK Foreign, Commonwealth and Development Office has created a digital diagnostic tool to conduct country diagnostics in three main areas: digital inclusion; capacity of government, society, and the economy to manage cyber risks; and status of the local digital economy. The resulting assessment and business case guide the office's digital programs.[120]

Adopting an established cybersecurity model allows better security risk management, produces cost saving, and supports good security procedures and processes. Despite these benefits, maturity models tend to provide a limited compliance assessment rather than a holistic approach to deal with emerging cyber environments, evolving usage patterns, as well as sophisticated attacks. To overcome these shortcomings, the assessment of cyber security maturity models needs to go beyond compliance checks, include quantitative metrics, and be sufficiently flexible to deal with emerging cyber risks.

## 4.4. The Next Generation of Cyber Risk Capabilities

Technology will continue its exponential growth path over the coming years. With 90% of the world's data having been collected in just the last 2 years, the volume and scope of data collections are set to accelerate. Smartphone manufacturers are on track to deliver 1.5 billion new units in 2022, while the number of IoT

116  World Bank and United Nations. 2017. *Combatting Cybercrime: Tools and Capacity Building for Emerging Economies.* Washington, DC: World Bank. https://doi.org/10.1596/30306.

117  Council of Registered Ethical Security Testers (CREST). Who Are CREST? https://www.crest-approved.org/about-us/who-are-crest/.

118  CREST. Cyber Security Incident Response Maturity Assessment. https://www.crest-approved.org/approved-services/cyber-security-incident-response-maturity-assessment/.

119  Organisation for Economic Co-operation and Development. 2022. OECD Policy Framework on Digital Security. Paris. https://doi.org/10.1787/a69df866-en.

120  Organisation for Economic Co-operation and Development. Leaving no one behind in a digital world: the United Kingdom's Digital Access Programme. https://www.oecd.org/development-cooperation-learning/practices/leaving-no-one-behind-in-a-digital-world-the-united-kingdom-s-digital-access-programme--e8b15982/.

devices installed globally is expected to reach 40 billion by the middle of the decade. Global internet penetration, at 62.5% in 2022, is projected to reach 94% by 2025. The outlook for AI is one of high growth and adoption.[121]

What impact will these trends have on cyber vulnerabilities and the next generation of cyber attacks (Table 9)? And how should organizations upgrade their cyber risk and defense capabilities? On-demand data access, adversarial AI attacks, and security-by-design represent three trends worth following over the coming years.[122]

**Table 9: The Next Generation of Cyber Attacks**

| Second Generation Cyber Attacks | Potential Strategic Applications |
|---|---|
| "Boomerang" Malware | "Captured" malware which can be turned back to operate against its owners |
| Weaponized Chatbots | Artificial Intelligence devices with enhanced capacities to deliver more persuasive phishing messages, quickly react to new events, and send message responses via social media; capable of attacking other bots |
| Deepfakes in a Cyber Physical War | Alterations to digital identity data (e.g., faces, voices, biomarkers), which are distorting the reality of what is occurring on the ground |
| Drone Swarms | Drones capable of hacking, disrupting communications (like Wi-Fi and Bluetooth), or engaging in surveillance |
| Quantum Computing | Devices with exponential computing power, able to break almost any encrypted system |

Source: M. McGuire. 2021. Nation States, Cyberconflict, and the Web of Profit. *HP Wolf Security* (blog). 8 April.

A first set of security issues is closely linked to the future of work which will increasingly be performed in remote and hybrid settings, made possible through on-demand access to data. This trend will require organizations to adopt new defensive cybersecurity capabilities such as the following:

- **Zero trust architecture.** In OECD countries, a quarter of all workers now work remotely 3 or more days a week; and globally, more than 60% of employees say they work remotely at least occasionally. The combination of hybrid and remote work and increased cloud access creates potential vulnerabilities. Under a zero trust architecture, cyber defenses would transition from traditional security models built around static perimeters (buildings, physical networks) to one that eliminates implicit trust and replaces it with continuous authorization at every stage of digital interactions. Zero trust models use strong authentication methods, leveraging network segmentation, preventing lateral movement, and simplifying "least access" policies. Organizations could tailor the adoption of zero trust capabilities to the threat and risk landscape they are

---

[121] A. McCain. 2022. How Fast Is Technology Advancing? [2022]: Growing, Evolving, and Accelerating at Exponential Rates. *Zippia Research*. April. https://www.zippia.com/advice/how-fast-is-technology-advancing/.

[122] McKinsey. Cybersecurity Trends: Looking over the Horizon.

facing and to their business objectives.[123] Access would be granted and enforced on a granular basis. Even if users have remote access to the data environment, they may be granted access to sensitive data only on a case-by-case, need-to-know basis.

- **Homomorphic encryption technologies** allow users to work with encrypted data without first having to decrypt the data, giving third parties and internal collaborators safer access to large data sets. With homomorphic encryption, the cloud service or outsourcing company has only access to encrypted data to perform computations before returning the encrypted results to the owner who can decrypt it with a private key. This approach enables organizations to share sensitive data in a secure and efficient manner without revealing the data or the results and helps companies comply with stringent data privacy requirements.[124]

- **Behavioral analytics.** Employees are a key vulnerability for organizations. Behavioral analysis monitors different aspects such as access requests or the security credentials of devices against a baseline to identify anomalous user behavior or device activity. These tools can enable risk-based authentication and authorization as well as orchestrate preventive and incident response measures.[125]

- **Elastic log monitoring** for large data sets offers enterprises a solution for keeping track of massive data sets that are being accessed by a highly decentralized workforce. Elastic log monitoring allows companies to pull log data from anywhere in the organization into a single location and to search, analyze, and visualize the data in real time.[126]

A second threat vector involves adversarial AI attacks. Cyber attackers will increasingly take advantage of sophisticated and intelligent technology solutions to find loopholes in IT systems. Cyber attacks that harness AI and machine learning may be the biggest threat facing organizations today. Adversarial AI attacks can, for instance, take the form of reverse engineering or the hijacking of AI models to access an AI system. A second AI attack type involves adversarial samples, whereby small sample instances introduce feature perturbations, causing AI models to learn from the manipulated data and therefore learn to classify incorrectly. As a result, the model becomes a source of incorrect classification (e.g., a self-driving vehicle that is designed to slow down at an intersection fails to recognize a manipulated stop sign). A third type of AI attack involves the manipulation of data that AI practitioners use to train the model. Once a criminal gains unauthorized access to the network, they can alter the data or introduce significantly different data sets.

To counter these sophisticated cyber attacks, organizations should institute a risk-based approach to automate their responses. For instance, organizations can automate lower-risk and labor-intensive processes (such as identity access management, software upgrades) and focus on higher-risk activities that require more direct oversight. To stay abreast of changing attack patterns, cybersecurity teams will need to scale up the use of defensive AI and machine learning to detect outlier patterns and noncompliance. Technical and operational changes may be required to strengthen defensive capabilities

[123] S. Rose. 2022. Planning for a Zero Trust Architecture: A Planning Guide for Federal Administrators. *NIST Whitepaper*. No. 20. https://doi.org/10.6028/NIST.CSWP.20.

[124] C. Dilmegani. 2021. What Is Homomorphic Encryption? Benefits & Challenges. *AIMultiple*. 19 August. https://research.aimultiple.com/homomorphic-encryption/.

[125] H. Ashtari. 2022. What Is Network Behavior Analysis? Definition, Importance, and Best Practices. *Spiceworks* (formerly Toolbox) (blog). 15 February. https://www.toolbox.com/tech/networking/articles/network-behavior-analysis/.

[126] Elastic. Log Monitoring with Elastic Observability. https://www.elastic.co/observability/log-monitoring.

against more sophisticated and frequent ransomware attacks. Technical changes include resilient data repositories and infrastructure, automated responses to malicious encryption, and advanced multifactor authentication to limit the potential impact of an attack. Organizational readiness can be improved by conducting tabletop exercises, developing detailed playbooks, and preparing for all contingencies to make the business response automatic.

A third line of defense against next-generation cyber attacks involves security-by-design. Increased regulatory scrutiny and gaps in response capacity will place new requirements on manufacturers to embed security into technology capabilities as they are being designed, built, and implemented.

- Rather than treating cybersecurity as an afterthought, organizations will need to embed it in the design of software from inception, including the use of a software bill of materials. One important way to create a secure software development life cycle is to have security and technology risk teams engage with developers throughout each stage of development. Another is to ensure that developers learn certain security capabilities themselves (for instance, threat modeling, code and infrastructure scanning, dynamic testing).[127]

- As compliance requirements grow, organizations can mitigate the administrative burden by formally detailing all components and supply chain relationships used in their software. This software bill of materials would list open-source and third-party components in a codebase through new software development processes, code-scanning tools, industry standards, and supply chain requirements.[128] In addition to mitigating supply chain risks, detailed software documentation helps ensure that security teams are prepared for regulatory inquiries.[129]

- Migrating workloads and infrastructure to third-party cloud environments (such as platform as a service, infrastructure as a service) can better secure organizational resources and simplify management for cyberteams. Cloud providers not only handle many routine security, patching, and maintenance activities, but also offer automation capabilities and scalable services. At the same time, cloud platforms have become magnets for hackers, who can cripple multiple firms with a single successful hack as illustrated in the SolarWinds example (Box 4, page 90).

- Standardizing infrastructure and control engineering processes can simplify the management of hybrid and multi-cloud environments and increase the system's resilience. This approach enables processes such as orchestrated patching, as well as rapid provisioning and deprovisioning.[130]

127  M. Souppaya, K. Scarfone, and D. Dodson. 2022. Secure Software Development Framework Version 1.1: Recommendations for Mitigating the Risk of Software Vulnerabilities. *NIST Special Publication*. 800-218. https://doi.org/10.6028/NIST.SP.800-218.

128  T. Bailey et al. 2022. Software Bill of Materials: Managing Software Cybersecurity Risks. *McKinsey*. 19 September. https://www.mckinsey.com/capabilities/risk-and-resilience/our-insights/cybersecurity/software-bill-of-materials-managing-software-cybersecurity-risks?stcr=56FBC168692C4D2AAA35085A62C05FD0&cid=other-eml-alt-mip-mck&hlkid=af755aefc5-9443ccaa4eb99eea038e9c&hctky=11257047&hdpid=35dcc0f7-4e98-4978-9e9c-e81e5af57cbe.

129  NIST. Software Security in Supply Chains: Software Bill of Materials. https://www.nist.gov/itl/executive-order-14028-improving-nations-cybersecurity/software-security-supply-chains-software-1.

130  T. Loehr. 2021. 8 Best Practices for Securing Infrastructure as Code. *Cycode* (blog). 6 October. https://cycode.com/blog/8-best-practices-for-securing-infrastructure-as-code/.

## 4.5. The Growing Importance of Cybersecurity for the Development Agenda

Until a few years ago, the niche nature of cybersecurity on the global policy agenda confined these issues to a small community of experts, mostly with a national security, law enforcement, or intelligence background, and stalled the integration of cyber capacity building into the development agenda.

Several factors help explain why the development community placed less attention to heightened cyber risks:[131]

- First, there is a knowledge gap. Development organizations initially did not consider traditional infrastructure projects as digital in nature and, therefore, failed to build in necessary de-risking measures to prevent cybercrime, unauthorized surveillance, disinformation, digital authoritarianism, or data exploitation that could end up undercutting development outcomes.
- A second factor is a communication challenge. Digital risks to projects can take many forms and typically have a technical underpinning. If these risks are not evaluated and communicated at the design stage, they will likely be left unaddressed throughout the project life cycle. As a result, organizations did not build the necessary de-risking processes and safeguards into their procurement, assistance, or investment operations.
- Third, understanding the specific needs of a country and knowing how to navigate high-stakes cyber issues require special skills and impartial advice. Development organizations often do not have the right implementors on hand, ranging from global consultancy firms to specialty boutiques, to international organizations (e.g., Organization for Security and Co-operation in Europe, ITU, UN Office on Drugs and Crime), to academic and nongovernment institutions (see Appendix 2 for an overview of specialized organizations and resources).
- A fourth obstacle is that cyber incidents are often politicized from a geopolitical perspective and entangled with data protection, trade, and competition issues. In that context, recipient countries may not fully appreciate the increased exposure to cyber risks and often lack the capacity to adapt cybersecurity preparedness to their specific requirements as opposed to simply copying laws and practices developed elsewhere.

Development practitioners are beginning to appreciate the vulnerabilities of development programs and projects to different types of cyber threats (Tables 8 and 9) and upgrade defensive cyber capabilities accordingly. For instance:

- Digital financial systems have become attractive targets for cyber attacks especially with payment modernization programs and projects which are revamping payment systems and processes built on over 20-year-old technology. Some notable payment modernization initiatives include

---

[131] W. van Pruisenweg. 2021. *Integrating Cyber Capacity into the Digital Development Agenda*. The Hague: Global Forum on Cyber Expertise.

open banking,[132] faster payments,[133] and the adoption of the ISO 20022 message.[134] A successful attack on a major financial institution, or on a core service used by many, could quickly spread through the entire financial system, causing widespread disruption and loss of confidence. Programming in digital financial services must include cybersecurity measures.

- Cybercriminals have used the turmoil of the COVID-19 pandemic as an opportunity to target strained health-care systems, crippling hospital computer systems and medical supply chains and extorting large payments. Recent reports describe ransomware as a national security threat with profound societal impacts and offer a comprehensive framework for action.[135]

Failure to address cyber risks may lower the public's trust in digital technologies. For development organizations, the message is clear: cyber incidents affecting development programs can cause long-term damage to their reputation and effectiveness in the partner country and around the globe.

Cybersecurity is a critical priority impacting development across all countries. Every sector that utilizes digital technology is affected by cybersecurity considerations:

- **Critical infrastructure sectors** such as energy, IT and communications, transportation, health, defense, finance, commerce, emergency services, water and dams, and chemical and nuclear installations underpin the functioning of any country.[136] These sectors use digital technologies to capture data, improve and expand service delivery, and optimize system performance. As these networks become more interconnected, securing critical infrastructure sectors is a prerequisite to the uninterrupted operation of essential services. A recent review by the US Government Accountability Office offers practical suggestions for identifying critical infrastructure priorities, seeking input from different stakeholders, improving coordination of cybersecurity services, and sharing threat information.[137]

- **Economic growth, finance, and trade.** The lack of cybersecurity is perceived as the single greatest threat to the global economy over the next decade (footnote 1). Institutions vital for economic and financial functioning of a country, like ministries of finance, central banks, and systemically important banks, are at risk of a cybersecurity breach. Vulnerabilities in digital technologies affect both large companies and SMEs and are common targets for cyber attacks,

---

[132] Open banking enables secure interoperability in the banking industry by allowing third-party service providers to access banking transactions and other data from banks and financial institutions to innovate financial services experience.

[133] Faster payments seeks to reduce the time it takes to process electronic payments to be able to complete payments and settlements almost immediately and at any time.

[134] ISO 20022 financial messages provide rich data in each transaction to automate manual and complex processes by eliminating the need for numerous messages required to complete a financial transaction. The standard messages also allow for structured communication of financial messages enabling more interoperability and connectivity between different parties and systems in the financial ecosystem.

[135] Institute for Security + Technology. 2022. *Combating Ransomware—A Comprehensive Framework for Action: Key Recommendations from the Ransomware Task Force.*

[136] The US Cybersecurity and Infrastructure Agency lists 16 critical infrastructure sectors whose assets, systems, and networks, whether physical or virtual, are considered so vital that their incapacitation or destruction would have a debilitating effect on national security, the economy, national public health or safety. Government of the US, Cybersecurity and Infrastructure Security Agency. Critical Infrastructure Sectors. https://www.cisa.gov/topics/critical-infrastructure-security-and-resilience/critical-infrastructure-sectors.

[137] Government of the US, Government Accountability Office. 2022. Critical Infrastructure Protection: CISA Should Improve Priority Setting, Stakeholder Involvement, and Threat Information Sharing. Press release. 1 March. https://www.gao.gov/products/gao-22-104279.

especially if they are part of a supply chain. The 2022 ransomware attack by the Russian Conti group on Costa Rica crippled key government agencies, most notably the finance ministry, whose online tax collection system went offline. This forced individuals and businesses to resort to paying in cash, causing major complications in people's lives and severely harming economic activity. The situation escalated to the point where, hours after his inauguration, President Chavez declared a state of emergency, seemingly the first time in history a national government has done so as a result of a cyber attack.[138]

- **Energy and environment.** According to industry surveys, more than half of gas, wind, water, and solar utilities around the world have reported at least one shutdown or operational data loss per year caused by cyber attacks.[139] As industrial systems digitize and increasingly connect globally, the digital attack surface is expanding. The World Economic Forum together with Siemens Energy and Saudi Aramco warned that unless the oil and gas value chain can withstand or recover rapidly from cyber risk, the risk may have disastrous physical, environmental, and safety consequences. Their blueprint draws on a set of general cybersecurity principles and proposes six resilience principles specific to the oil and gas industry (Figure 14).[140]

**Figure 14: Cyber Resilience Principles for the Oil and Gas Industry**

**10 General Cyber Resilience Principles**

1 Responsiblity for cyber resilience
2 Command of the subject
3 Accountable officer
4 Integration of cyber resilience
5 Risk appetite
6 Risk assessment and reporting
7 Resilience plans
8 Community
9 Review
10 Effectiveness

**+**

**6 Oil and Gas Industry-Specific Cyber Resilience Principles**

1 Cyber Resilience governance
2 Resilience by design
3 Corporate responsibility for cyber resiliene
4 Holistic risk-management approach
5 Ecosystem-wide collaboration
6 Ecosystem-wide cyber resilience plans

Source: F. Beato et al. 2021. Cyber Resilience in the Oil and Gas Industry: Playbook for Boards and Corporate Officers. World Economic Forum, Siemens Energy, and Saudi Aramco. 17 May.

---

138  S. Lichtenstein. When a Country Is Held Hostage: What to Make of Costa Rica's Ransomware Attack. *Risk Assistance Network Exchnge-Stratfor*. 26 May. https://worldview.stratfor.com/article/article/when-country-held-hostage-what-make-costa-ricas-ransomware-attack.

139  Siemens. 2019. Siemens and Ponemon Institute study finds utility industry vulnerable to cyberattacks. Press release. 4 October. https://press.siemens.com/global/en/pressrelease/siemens-and-ponemon-institute-study-finds-utility-industry-vulnerabilities.

140  F. Beato et al. 2021. Cyber Resilience in the Oil and Gas Industry: Playbook for Boards and Corporate Officers. World Economic Forum, Siemens Energy, and Saudi Aramco. 17 May. https://www.weforum.org/whitepapers/cyber-resilience-in-the-oil-and-gas-industry-playbook-for-boards-and-corporate-officers/.

■ **Food security and agriculture.** Digital technologies will play an increasing role in boosting agricultural productivity to meet global food demand. The move toward precision agriculture is accelerating. As automated machinery, high resolution imagery, drones, soil sensors, and a range of other IoT technologies are expanding, farms will generate an ever greater amount of agricultural data. Market trends suggest that a small number of companies will produce most agricultural technology devices, rendering them highly vulnerable to cyber attacks and data theft. In a 2020 review, researchers presented the implications of cyber attacks and cyberterrorism (on data, network and equipment, supply chains, and cloud computing) as smart devices make it possible to remotely control and exploit field sensors and autonomous vehicles. This would provide the ability to destroy crops by flooding farmlands, overspraying pesticides, or causing disruption of cold storage facilities, ultimately causing unsafe consumption or food shortages.[141]

The food and agriculture sectors lack cyber threat awareness, which has led to a discrepancy between the expansion of digital agriculture technologies and necessary upgrades to digital security. Cyber attacks on this very connected industry can have repercussions for farmers, food processing industries, agriculture cooperatives, distributors, consumers, government agencies, and nations dependent on income from agriculture exports. The two areas that enable digital farming—connectivity and information flow—represent the highest vulnerabilities and should be protected first.[142] In addition, this sector is closely linked with many other critical infrastructure sectors, including water and wastewater systems, transportation systems, the energy sector, and the chemical sector. Any disruption to the food and agriculture sectors will have knock-on effects on other critical infrastructure sectors. As such, it is of vital importance to protect their operations.

■ **Education.** Although global data of cyber attacks on the educational sector is limited, academic institutions which host a wealth of data, including personal information and sensitive research are a potential target. Universities are repositories of valuable proprietary and personal data, often tied to corporations and government organizations that span medical, military, financial and emerging technologies, and more.[143] Attacks on educational institutions are growing faster than on any other sector. A recent survey found that nearly half of all education institutions globally were targeted by ransomware in 2020, with 58% of those saying that cybercriminals succeeded in encrypting their data. Educational institutions lost $2.73 million in an average ransomware incident, nearly $300,000 more than distributors and transportation companies, the next highest-ranked sector.[144] As was evident during the COVID-19 pandemic, research universities are most threatened by state-sponsored hacking and cybercriminals looking to sell stolen data to nation-state actors.

■ **Health.** Technology is shaping the future of health care across the world. However, health care is also one of the sectors most fraught with cybersecurity concerns. Cybercriminals target the sector because of its many vulnerabilities, the high black market value of its data, and the

[141] M. Gupta et al. 2020. Security and Privacy in Smart Farming: Challenges and Opportunities. *IEEE Access.* Volume 8: 34564–84, https://doi.org/10.1109/ACCESS.2020.2975142.

[142] J. Ducker. 2022. Investigating the Vulnerability of the Food Industry to Cyberattacks. *AZO Life Sciences.* 4 March. https://www.azolifesciences.com/article/Investigating-the-Vulnerability-of-the-Food-Industry-to-Cyberattacks.aspx.

[143] A. Huls. 2021. To Improve Higher Ed Data Security, Address These Risks in Research Projects. *EdTech Magazine.* 10 May. https://edtechmagazine.com/higher/article/2021/05/improve-higher-ed-data-security-address-these-risks-research-projects-perfcon.

[144] B. Freed. 2021. 44% of education institutions targeted by ransomware in 2020, survey finds. *EdScoop.* 23 July. https://edscoop.com/ransomware-education-institutions-sophos/.

willingness of many health-care providers to pay ransom in order to recover access to their compromised systems and data. The health sector's commitment to interoperability and open-source technologies may actually increase the likelihood of a cyber incident. The 2017 WannaCry ransomware virus is perhaps the most well-known, crippling the UK's national health-care system for a week. A June 2020 cyber attack on South Africa's largest private hospital operator, Life Healthcare Group, affected its admissions systems, business processing, and email servers as the hospital struggled to meet the influx of patients seeking treatment for COVID-19 symptoms. The growth of online misinformation/disinformation in the health sector poses an additional threat to public health.

■ **Human rights and governance** programs often work in areas considered to be politically sensitive, including with marginalized groups, refugees, and human trafficking. Cyber surveillance, cyber attacks, and the exposure of personally identifiable information (PII) are increasingly used to disrupt or manipulate these programs and threaten their beneficiaries.[145] Mobile social platforms and messaging apps are generating massive amounts of personal data whose detailed user profiles can be used to influence opinions and exploit social divisions. A sophisticated cybersecurity attack against computer servers hosting information held by the International Committee of the Red Cross compromised personal data and confidential information on more than 515,000 highly vulnerable people, including those separated from their families due to conflict, migration, and disaster; missing persons and their families; and people in detention.

An important first step to bridge this gap between development and cybersecurity communities is to re-frame the narrative for cybersecurity capacity building in terms of digital resilience, safety, trust, sustainability, and risk management and to link security with sustainable economic development and digital human rights.

■ As an example, the "Boe Declaration on Regional Security," which was adopted by the leaders of Pacific Islands Nations in 2018, offers a blueprint and action plan for addressing cybersecurity as a crosscutting priority as well as a development priority for increasing investment and collaboration across the region.[146]

■ Organized social media campaigns have become a growing venue for spreading misinformation/disinformation across the globe. In 2020, these types of campaigns were documented in 81 countries, up from 28 in 2017,[147] using social media platforms to influence events in other countries and to destabilize. These campaigns increasingly use legally or illicitly acquired data to monitor and analyze target populations, produce fake information or AI-generated images to polarize political and social dynamics, and undermine trust. To counteract fake news, the United Nations Development Programme's automated fact-checking tool iVerify was added to the Digital Public Goods (DPG) Registry of the Digital Public Goods Alliance as the first digital public good for fighting misinformation.[148] It is meant to help combat the spread of false

[145] International Committee of the Red Cross. 2022. Sophisticated Cyber-Attack Targets Red Cross Red Crescent Data on 500,000 People. News release. 19 January. https://www.icrc.org/en/document/sophisticated-cyber-attack-targets-red-cross-red-crescent-data-500000-people.

[146] Pacific Islands Forum. BOE Declaration Action Plan. https://www.forumsec.org/wp-content/uploads/2019/10/BOE-document-Action-Plan.pdf.

[147] *The Economist.* 2021. A Growing Number of Governments Are Spreading Disinformation Online. News release. 13 January. https://www.economist.com/graphic-detail/2021/01/13/a-growing-number-of-governments-are-spreading-disinformation-online.

[148] United Nations Development Programme (UNDP). 2022. UNDP Tool to Fight Misinformation Scales Globally as a Digital Public Good. Press release. 17 May. https://www.undp.org/press-releases/undp-tool-fight-misinformation-scales-globally-digital-public-good.

narratives during elections. The United Nations Development Programme also published a field-tested tool kit to help youth counter fake news and play a pivotal role in verifying news.[149]

# 4.6. Integrating Cybersecurity into Development Programs

The shared goal should be to achieve an open, secure, and inclusive digital ecosystem that safeguards a country's critical infrastructure and economic activities, facilitates internet freedoms, ensures the free flow of data, protects intellectual property, and meets countries' cybersecurity needs.

Recent initiatives and reports call for the integration of cybersecurity and cyber capacity into digital development programs as an opportunity to achieve better outcomes by building strong resilience, safety, security, and trust into digital transformation efforts.[150] As discussed in the previous section, the protection of critical infrastructure sectors has become an important part of the national agenda in many countries, requiring the engagement of the highest levels of government and industry.

As these integration efforts are taking shape across the development community, emerging practices can be summarized along five levels: (i) organizational integration, (ii) country diagnostics, (iii) systemic interventions, (iv) partnership initiatives, and (v) sector applications (see Appendix 3 for an overview of cybersecurity resource material).

At an organizational level, USAID's *Cybersecurity Primer* offers guidance on how to incorporate a risk-based approach into country strategies, program and project design and implementation, and adaptive, learning-oriented feedback loops (Table 10).

Other development organizations are pursuing similar adaptations. The European Bank for Reconstruction and Development, for instance, is piloting advance (ex ante) reviews to assess digital opportunities and risks as part of its transition approach to competitive, inclusive, green, well-governed, integrated, and resilient economies and is using digital maturity road maps and cybersecurity audits to inform remedial action and technology selection (Box 3). The World Bank, for its part, has a capacity development program on Cybersecurity and Resilience of Critical Infrastructure in partnership with Korea Internet and Security Agency and the Global Knowledge Exchange and Development Center.[151]

---

[149] UNDP. 2021. Fake News and Social Stability. https://www.undp.org/lebanon/publications/fake-news-and-social-stability.

[150] For examples, see the following: *Xinhua News Agency*. 2021. G20 Ministers Agree on Digital Working Group, Discuss Cybersecurity. 6 August. http://www.xinhuanet.com/english/2021 08/06/c_1310110572.htm; B20 Italy. Final Communiqué: Policy Recommendations to the G20. https://www.b20italy2021.org/wp-content/uploads/2021/10/B20_Final-Communique_Web.pdf; European Commission. The EU Cybersecurity Act. https://digital-strategy.ec.europa.eu/en/policies/cybersecurity-act; OECD. Encouraging Vulnerability Treatment: Overview for Policy Makers. https://www.oecd.org/digital/encouraging-vulnerability-treatment-0e2615ba-en.htm; UN Office of the Secretary-General's Envoy on Technology. 2020. Report of the Secretary-General: Roadmap for Digital Cooperation. www.un.org/techenvoy/sites/www.un.org.techenvoy/files/general/Roadmap_for_Digital_Cooperation_9June.pdf; Global Forum on Cyber Expertise. News. https://thegfce.org/news-articles; USAID. 2021. *Cybersecurity Primer*. www.usaid.gov/digital-development/usaid-cybersecurity-primer; The European Bank for Reconstruction and Development. The EBRD's Digital Approach. https://www.ebrd.com/ebrd-digital-approach.html; and Ransomware Task Force. 2021. Combating Ransomware: A Comprehensive Framework for Action. Institute for Security and Technology. https://securityandtechnology.org/ransomwaretaskforce/report/.

[151] World Bank: Supporting Countries in Building Cybersecurity and Resilience of Critical Infrastructure. https://www.worldbank.org/en/programs/kodi/brief/supporting-countries-in-building-cybersecurity-and-resilience-of-critical-infrastructure.

**Table 10: Integrating Cybersecurity into Development Strategies, Programs, and Projects**

| Country/Regional Strategic Planning | Project Design and Implementation | Activity Design and Implementation | Monitoring, Evaluating, Learning |
|---|---|---|---|
| • Assess partner country policies and strategies for cybersecurity, data privacy, communications infrastructure, and digital tools to identify areas of engagement<br>• Seek inputs and learn from local ICT and cybersecurity stakeholders to understand digital ecosystem<br>• Identify areas of alignment<br>• Coordinate with other donors to identify common areas<br>• Include cybersecurity concerns in the risk analysis and mitigation plan for each country strategy<br>• Incorporate crosscutting themes about cyber threats, disinformation, or online harassment into the results framework | • Research how key actors within a sector use technology, what systems they use and for what, as well as typical disinformation pathways to better identify project-specific cyber vulnerabilities<br>• Assess the importance of digital trust for a project's success<br>• Embed cybersecurity resources and responsibilities into every project at the design stage (draw on digital ecosystem country assessments; include in the key risk section of project documents);<br>• Develop stand-alone cybersecurity elements within projects with digital components | • Design and procure activities that support the development of resilient cybersecurity in partner countries<br>• Protect digital tools/solutions by installing licensed hardware and software; support interoperability and adopt cybersecurity standards; promote digital literacy; include budget resources to procure and sustain digital tools<br>• Build trust into activity design and implementation (e.g., engage local technical working group; identify trusted digital tools; establish plans for quality assurance and change management)<br>• Find creative, low-cost ways of incorporating cybersecurity into implementation meetings. | • Develop cybersecurity-related indicators to measure the effectiveness of cybersecurity interventions (e.g., indicators to assess upskilling and cybersecurity capacity building)<br>• Compile lessons from implementing cybersecurity interventions to scale successful interventions |

ICT - information and communication technology.
Source: USAID. 2021. Cybersecurity Primer.

---

**Box 3: Implementing EBRD's First Digital Approach**

Complementing the theme of *Digital Dividends* of its 2021–2022 Transition Report, the European Bank for Reconstruction and Development (EBRD) announced a comprehensive framework to mainstream digital technologies throughout the bank's activities, notably for addressing climate change, building a net-zero economy, and enabling smart green cities. As part of this initiative, its Digital Hub is preparing a cybersecurity tool kit to help clients with maturity assessments, penetration testing, and technical skill review, which is being tested for small and medium-sized enterprises' operations and smart city initiatives.

EBRD's Impact Team is tagging digital project components to identify areas where digital operations can have the most impact. In parallel, discussions are underway on the criteria for deciding which operations require an in-depth digital risk assessment, with a preference for maintaining a light-touch process.

Sector units are developing technology compendiums (for water, building and energy efficiency, transport, energy, and solid waste) to take stock of available technology packages and identify digital maturity road maps, ranging from basic analog to advanced digital. This assessment will influence upstream design considerations, notably whether proposed digital components are appropriate given the country's cybersecurity and capacity building. Sector teams are using a short questionnaire (5–10 items) to assess impact on critical infrastructure, staffing, preparedness, institutional setup, etc. This assessment may be followed up with a third-party risk audit to identify remedial cybersecurity actions that the company has to put in place by a date certain for the operation to be green-lighted.

Source: A. Reiserer. 2021. EBRD adopts First Digital Approach. *EBRD*. 10 November. Annual Review 2022.

At the country level, a baseline assessment can provide valuable insights into the current cybersecurity landscape and maturity level, identify key actors in the cybersecurity space, and gauge the perceptions of different stakeholder groups regarding the adequacy of the policy, regulatory, and legal environment for cybersecurity (Table 11). This step can also set the stage for future engagements by addressing questions such as the following:

- What are the cyber risks and opportunities in the country's digital ecosystem?
- How could these risks affect development outcomes?
- What mitigation measures need to be integrated at strategic and operational levels?

### Table 11: Assessing the Cybersecurity Landscape, Risks, and Opportunities

| Current State and Impact | |
| --- | --- |
| • What information is publicly available on the state of cyber threats and trends, particularly through government and private sector analysis and reporting? How might these threats and trends affect the programming?<br>• Are the programs involved in any activities that might be perceived as politically sensitive (e.g., advocacy for democratic norms, human rights, or anti-corruption) that may increase the likelihood of attacks and data breaches?<br>• Do the programs utilize and depend on any hardware or software that is vulnerable to attacks? Have these technologies been designed from the start with security as a priority? Is the software used by partners and/or beneficiaries licensed and current?<br>• What cybersecurity policies, regulations, and legislation exist in the country? To what extent are they implemented?<br>• What, if any, organizations exist that raise awareness of cybersecurity risks? How do they raise awareness? What impact to they have on increasing cybersecurity practices? | • What cybersecurity measures are in place to protect critical internet infrastructure?<br>• What is the extent of different actors' (user, business, government) capability to understand and use cybersecurity products and standard practices?<br>• Do higher education institutions offer curricula on cybersecurity? Are these programs adequate to prepare current and future demand for information security workforce skills?<br>• What does the competitive landscape look like in terms of cybersecurity providers (e.g., many vs. few; local vs. international companies)?<br>• Have there been any recent high-profile data breaches or cybersecurity incidents (private or public sector)? At what scale? How were they handled? Was there any communication issued by the government? By the private sector?<br>• Do institutions or organizations undergo information audits (such as penetration testing) to ensure the validity of cybersecurity strategies and policies in place?<br>• Do cybersecurity standards limit the growth of tech startups and small and medium-sized enterprises? |
| **Cybersecurity Actors** | **Perceptions** |
| • Who are the key cyber threat actors related to your environment and programs (e.g., criminal entities, domestic actors, foreign nation states, businesses, political parties, hacktivists, etc.)?<br>• What stakeholders are engaged in policymaking, advocacy, or programming on cybersecurity (e.g., civil society organizations, tech companies, government ministries, donors)?<br>• What stakeholders are responsible for monitoring and enforcing cybersecurity threats?<br>• Is there a national cyber/critical incident response team, computer emergency response team, or a computer security incident response team? Which individuals or organizations are members? What is its mandate? | • How do different stakeholders (civil society, private sector, government, individuals) perceive the importance of cybersecurity?<br>• How is the policy, regulatory, and legal environment for cybersecurity perceived by different stakeholders (individuals, private sector, business associations, civil society organizations)?<br>• What is the perception by different stakeholders of government's capacity to monitor, detect and react to cybersecurity breaches?<br>• How do people's concerns around cybersecurity threats affect their online activities?<br>• To what degree do cybersecurity concerns deter investment in new technologies? |

Source: USAID. 2021. *Cybersecurity Primer.*

To lay the groundwork for a system-level approach on cybersecurity, either through a stand-alone cybersecurity component or by integrating cybersecurity across a program, potential interventions can apply a people, process, and technology lens (Table 12). The "people" dimension is the most critical aspect as it examines the capacity, behavior, and resources that an individual within an organization is expected to maintain to remain secure in any digital activity they carry out. The "process" dimension is focused on procuring appropriate technology, response protocols for cyber attacks, and approaches for knowledge sharing. The "technology" lens is focused on sector-specific risk assessments and organizational maturity to determine appropriate and cost-effective technology solutions to keep an organization's digital ecosystem secure.

**Table 12:** Applying System-Level Approaches to Cybersecurity Interventions

| People | Processes | Technology |
|---|---|---|
| • Build in-house capacity of team members to understand why cybersecurity and its societal implications are relevant for programs across sectors. [...]<br>• Institute digital literacy and cyber hygiene training across all levels.<br>• Build a robust pipeline of cybersecurity and technology professionals in the country, by working closely with appropriate institutions and/or with other donors.<br>• Leverage programs to support the local technology ecosystems that can then provide local talent to solve local cybersecurity challenges.<br>• Support the development of a cyber-resilient civil society that monitors digital trends in a country, advocates for open, secure, and interoperable digital systems, and educates the population on cyber threats and security. | • Conduct a digital ecosystem country assessment to understand the digital context of the partner country, including legislation related to cybersecurity. [...]<br>• Build trust initiatives that encourage local processes or frameworks for NGOs to share information on observed cybersecurity attacks.<br>• Encourage partners to adopt risk-based approaches to cybersecurity, including clear processes to identify, protect, detect, respond to, and recover from cyber attacks. Such processes tend to be resource intensive.<br>• When designing programs, incorporate lessons learned from existing cybersecurity programs.<br>• Establish a cybersecurity donor coordination group in-country to discuss cyber challenges and opportunities.<br>• Across programming, encourage information sharing related to cyberattacks or cybersecurity vulnerabilities. | • Consider conducting an assessment of potential risks associated with proposed digital solutions for your sector by an implementing partner or outside experts for a second opinion.<br>• Encourage partners to understand the VPN marketplace, including whether their use is permitted or whether they may be monitored, before using them.<br>• Encourage the transparent procurement of software and hardware technology assets in-country to prevent or limit the use of pirated digital assets that may be compromised. |

NGOs = nongovernment organizations, VPN = virtual private network.
Source: USAID. 2021. *Cybersecurity Primer.*

At a partnership level, a variety of initiatives demonstrate growing collaboration between cyber capacity efforts and digital development. The following is a sample of initiatives:

■ The EU, Council of Europe, and Interpol launched its Global Action on Cybercrime Extended. Operating with hubs in 15 countries, the project seeks to strengthen the capacities of countries to apply legislation on cybercrime and electronic evidence, enhance the abilities of authorities to investigate and prosecute cybercrime, and engage in effective international cooperation in this area.[152]

---

152  Council of Europe. Global Action on Cybercrime Extended (GLACY)+. https://www.coe.int/en/web/cybercrime/glacyplus.

- The Association of Southeast Asian Nations has been working toward establishing a regional Cyber Emergency-Response Framework[153] since 2016, with informal bilateral collaboration between computer emergency response teams since the early 2000s, a Coordinating Committee on Cybersecurity and several agreed strategy documents, including the Cybersecurity Cooperation Strategy, 2021–2025.[154]
- The Pacific Cyber Security Operational Network was established in 2018 with Australian government aid funding and backing by Australia's International Cyber Engagement Strategy. It is a peer-to-peer knowledge and information sharing network of government-designated cybersecurity incident response officials comprising of the following members: Australia, the Cook Islands, Fiji, Kiribati, the Marshall Islands, Nauru, New Zealand, Niue, Palau, Papua New Guinea, Samoa, the Solomon Islands, Tokelau, Tonga, Tuvalu, and Vanuatu.[155]
- The EU Cybernet Platform fields vetted EU experts from the cybersecurity community to increase cyber resilience and capacities, provide technical assistance to tackle malicious cyber activities, and strengthen the delivery of the EU's external cyber capacity-building projects. [156]
- The Council of Europe has established a 24/7 network to combat cybercrime under the Budapest Convention on Cybercrime. Parties to the convention can send and receive requests for assistance and provide alerts on cyber threats to critical infrastructures in other countries.[157]
- The Organization for Security and Co-operation in Europe's Point-of-Contact Network of policy and technical contact points serves as a regional risk reduction mechanism aimed at preventing conflicts stemming from the misuse of information and communication technologies by nation states.[158]
- USAID's Digital APEX program provides teams and partners with a roster of pre-approved cybersecurity services and experts to receive rapid technical assistance to preemptively reduce digital vulnerabilities, mitigate damages from malicious acts, conduct digital risk assessments, secure digital transactions, or assist with software and hardware procurement, cyber training, and incident response.[159] USAID also funds multiple projects with a cyber security focus under its Digital Connectivity and Cybersecurity Partnership,[160] including in East Asia and Pacific, Central and South Asia.
- CREST's Service Selection Platform lists vetted vendors who offer free guidance to governments, regulators, and buyers on technical security services and procurement of information and communication technology and services.[161]

153  K. L. Tay. 2023. ASEAN Cyber-security Cooperation: Towards a Regional Emergency-response Framework. *International Institute for Strategic Studies.* 22 June. https://www.iiss.org/research-paper/2023/06/asean-cyber-security-cooperation-towards-a-regional-emergency-response-framework/.

154  ASEAN. 2021. ASEAN Cybersecurity Cooperation Strategy 2021–2025. Jakarta: Association of Southeast Asian Nations. https://asean.org/wp-content/uploads/2022/02/01-ASEAN-Cybersecurity-Cooperation-Paper-2021-2025_final-23-0122.pdf.

155  Pacific Cyber Security Operational Network. https://pacson.org/.

156  EU CyberNet. https://www.eucybernet.eu/.

157  Council of Europe. The 24/7 Network Established under the Budapest Convention on Cybercrime. https://www.coe.int/en/web/cybercrime/24/7-network-new-.

158  Organization for Security and Co-operation in Europe. OSCE Cyber Security Awareness Month. https://www.osce.org/cyber-security-awareness-month.

159  Bixal. Building Cybersecurity Capacity of Civil Society Organizations in Colombia to Improve Digital Health and Protect against Cyber Threats. https://www.bixal.com/case-studies/digital-apex/.

160  USAID. Digital Connectivity and Cybersecurity Partnership. https://www.usaid.gov/digital-development/digital-connectivity-cybersecurity-partnership.

161  CREST. Member Companies. https://service-selection-platform.crest-approved.org/accredited_companies/regions_and_services/.

- The Organization for American States and the EU Cyber Resilience for Development Unit conduct online training and tabletop exercises for incident responders to test maturity and identify vulnerabilities.[162]
- The International Finance Corporation (IFC) has incorporated de-risking mechanisms through its financing. The risk management framework helps assess risks for specific industry sectors, with those considered to have higher risk exposure (e.g., energy, logistics) receiving additional scrutiny versus sectors considered lower risk (e.g., finance, telecom) because of existing industry frameworks, best practices, and regulations.

---

[162] Cyber Resilience for Development. We are Cyber4Dev. https://cyber4dev.eu/.

**Strategic approach.** Implementing
system-level approaches for robust
cybersecurity interventions
(photo by ADB).

# 5

## THIRD-PARTY DIGITAL RISK MANAGEMENT

The growing presence of third-party vendors and service providers in supply chains is impacting organizations' exposure to new types of digital risks.[163] These third-party digital risks are on the rise, as shown by the record numbers of third-party data breaches (e.g., Log4j,[164] PracticeMax,[165] Kaseya,[166] SolarWinds,[167] and many others); supply chain disruptions from cyber breaches (e.g., Toyota);[168] ongoing pandemic shutdowns; and the Russian invasion of Ukraine.[169] Gartner predicts that by 2025, 45% of organizations worldwide will have experienced attacks on their software supply chains—a three-fold increase from 2021.[170]

The purpose of third-party risk management (TPRM) is to analyze and mitigate these risks and anticipate new trends on the horizon that deserve attention.

Over three-quarters of organizations are increasing the number of third parties they work with to accomplish their mission-critical priorities through collaboration with or outsourcing to third parties.[171] These arrangements often deliver superior deployment in terms of speed, efficiency, and adaptability by shortening production or delivery cycles as well as providing access to innovative ideas. It is often better to outsource expertise for highly specialized tasks related to digital technologies, particularly when there is no in-house capacity or if it is considered insufficient.

## 5.1. The Complex Risk Landscape of Third-Party Collaboration

Third-party collaboration raises new types of risks that cut across several domains, including cyber, operational, financial, environmental, reputational, and geopolitical risks, and increases the complexity of information security. Some third parties, including startups and other innovators, might be performing new-in-kind technology and analytics services, providing services outside of the company's core business model which may include prototyping and testing and may be keen to evolve their product or service based on the derived insights. Such engagements demand a fundamentally different approach to risk identification and monitoring.

---

163  A third party includes all entities that have entered into business relationships with an organization. This third-party ecosystem includes, among others, contractors, partners, joint ventures, service providers, intermediaries, agents, suppliers, and consultants.

164  M. Hill. 2022. The Apache Log4j Vulnerabilities: A Timeline. *CSO Online*. 7 January. https://www.csoonline.com/article/3645431/the-apache-log4j-vulnerabilities-a-timeline.html.

165  E. Shaak. 2022. PracticeMax Facing Class Action Over 2021 Data Breach Affecting 150K Patients. *ClassAction*. 28 July. https://www.classaction.org/news/practicemax-facing-class-action-over-2021-data-breach-affecting-150k-patients.

166  C. Osborne. 2021. Updated Kaseya Ransomware Attack FAQ: What We Know Now. *ZDNET*. 23 July. https://www.zdnet.com/article/updated-kaseya-ransomware-attack-faq-what-we-know-now/.

167  S. Oladimeji. 2022. SolarWinds Hack Explained: Everything You Need to Know. *TechTarget*. 29 June. https://www.techtarget.com/whatis/feature/SolarWinds-hack-explained-Everything-you-need-to-know.

168  S. Zurier. 2022. Code Vulnerability Failures in Manufacturing on Display in Toyota Supply Chain Attack. SC Magazine. 1 March. https://www.scmagazine.com/analysis/supply-chain/code-vulnerability-failures-in-manufacturing-on-display-in-toyota-supply-chain-attack.

169  Cyber Peace Institute. Cyber Attacks in Times of Conflict. https://cyberconflicts.cyberpeaceinstitute.org/.; and B. Smith. 2022. Defending Ukraine: Early Lessons from the Cyber War. *Microsoft on the Issues* (blog). 22 June. https://blogs.microsoft.com/on-the-issues/2022/06/22/defending-ukraine-early-lessons-from-the-cyber-war/.

170  S. Moore. 2022. 7 Top Trends in Cybersecurity for 2022. *Gartner*. 13 April. https://www.gartner.com/en/articles/7-top-trends-in-cybersecurity-for-2022.

171  Gartner. Third-Party Risk Management Governance and Technology Investments. https://www.gartner.com/en/legal-compliance/trends/third-party-risk-governance-and-technology.

"Normal" disruptions to business operations caused by third parties such as supply shortages or price increases are typically handled via contractual agreements, business continuity plans, and incidence response procedures. Yet, the sheer number of direct and indirect third-party connections (Figure 15) in digital technologies multiplies the potential attack vectors and presents a major challenge for organizations trying to understand the full extent of their third-party relationships and the underlying risks.[172] Organizations do not have full control over third parties nor do they have complete transparency into their security protocols. Not all vendors have robust security standards and good risk management practices. A well-known example of this occurred when one of Target's heating, ventilation, and air-conditioning contractors inadvertently exposed millions of credit cards as a result of a hacking incident.[173]

Digital risks within this category include exposure or loss due to a cyber attack or security breach, because third parties had been granted too much privileged access to intellectual property, sensitive data, personally identifiable information (PII), or protected health information. Cloud adoption accounts for a growing list of data exposures, such as the Cloud Hopper attack.[174] This attack, which affected organizations worldwide, compromised several IT service providers that were used as intermediaries to infiltrate the networks of

**Figure 15: The Increasingly Extended Enterprise and its Third Party Network**

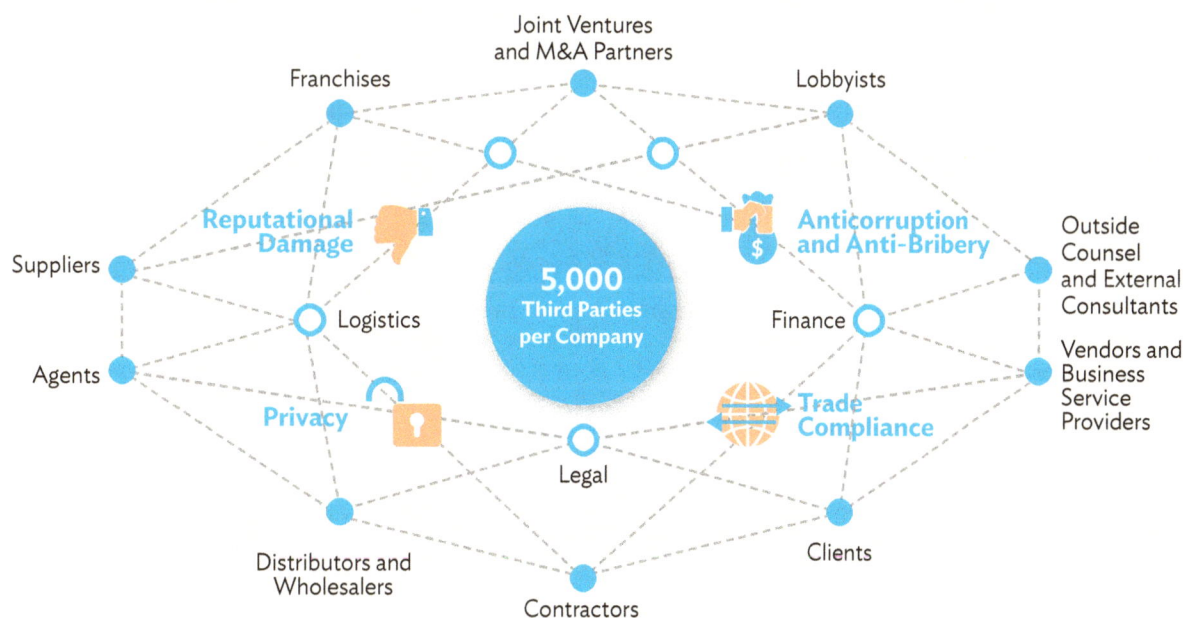

M&A = mergers and acquisitions.
Source: D. Hebda. Maximize Third-Party Risk Management with Aligned Assurance. Gartner (webinar). On-demand

172   The average number of parties with whom an enterprise shares sensitive information is 583. Security Scorecard. How to Manage Third-Party Digital Risk. https://securityscorecard.pathfactory.com/all/how-to-manage-third-party-risk.
173   B. Krebs. 2014. Target Hackers Broke in Via HVAC Company. Krebs on Security. 5 February. https://krebsonsecurity.com/2014/02/target-hackers-broke-in-via-hvac-company/.
174   R. Barry and D. Volz. 2019. Ghosts in the Clouds: Inside China's Major Corporate Hack. The Wall Street Journal. 30 December. https://www.wsj.com/articles/ghosts-in-the-clouds-inside-chinas-major-corporate-hack-11577729061.

the real targets to obtain sensitive corporate data. Mega breaches have followed a methodical "hack one, breach many" approach that starts by attacking a third-party vendor that serves multiple organizations and enterprises (Box 4).

Legislation, such as Australia's Foreign Influence Transparency Scheme Act, the US Foreign Corrupt Practices Act, and the UK Bribery Act, reinforce the principle that organizations are accountable for the decision to engage with a third party. Data protection laws and regulations, like the EU's General Data Protection Regulation (GDPR),[175] the California Consumer Privacy Act (CCPA),[176] Canada's Personal Information

---

## Box 4: The SolarWinds Supply Chain Attack and Its Aftermath

On 13 December 2020, FireEye, a global cybersecurity firm, posted the first details about the SolarWinds supply chain attack—the largest and most sophisticated attack to date. This global intrusion campaign inserted a trojan in the SolarWind's Orion business software update to distribute a malware, named "Sunburst." The attackers added a configuration file to the Orion product, which had been digitally signed and enabled backdoor communications over normal, unencrypted web traffic to other servers. The Sunburst malware is suspected to have lain quietly for several weeks while it performed some reconnaissance that led to file transfers and finally the control of the victim's servers. Sunburst blended in with Orion's normal functions and even had the ability to detect antivirus and cybersecurity forensic tools.

The attack was a coup for the suspected perpetrators, thought to be a nation-state actor, who had potential access to as many as 18,000 SolarWinds customers. It was used to steal valuable intellectual property from FireEye and infiltrated dozens of United States (US) cabinet-level agencies (Treasury, Justice, State Department, Commerce); academic institutions; nongovernment organizations; and technology companies. Due to the pervasiveness of the SolarWinds product across the world, more breaches will be found, while others may never be discovered or admitted.

Brad Smith, president of Microsoft, offered the following assessment in December 2020: "This is not 'espionage as usual,' even in the digital age. Instead, it represents an act of recklessness that created a serious technological vulnerability for the United States and the world. [It] also provides a powerful reminder that people in virtually every country are at risk and need protection irrespective of the governments they live under."

SolarWinds was clearly not performing its own due diligence and due care to protect itself and its customers. At the same time, SolarWinds' customers were also not following best practice. If they had performed their own cybersecurity assessment on this third-party software maker, this attack could have been detected. Were intake questions asked about the type of data to which SolarWinds had access and where that data might be stored? What was SolarWind's supply chain security? Were any external scans performed? Had SolarWinds remediated the findings of the external security scans?

Across the security industry, the SolarWinds hack served as a wake-up call for demanding greater insights into the provenance and integrity of software. Keeping tabs on proprietary systems like Orion is challenging because security tools need to foster transparency without exposing intellectual property. The problem is especially acute for open-source software, where projects may not have stable funding—which can lead to a lack of maintenance and updates. In addition, developers often repurpose useful parts of open-source code, which in turn means that a supply chain attack can compromise an open-source tool and end up spreading malicious updates across multiple applications.

In the wake of the SolarWinds attack, the US government issued an executive order that addressed numerous cybersecurity aspects, with requirements for federal agencies to generate guidelines, conduct evaluations, and implement improvements, such as software bills of materials. Google announced $10 billion in security investment over 5 years, with software supply chains as a high priority. Github is working on systems to automatically spot vulnerabilities in open-source projects. And the Open Source Security Foundation released a set of open source best practices in August 2022 to achieve supply chain security.

Source: G. Rasner. 2021. Cybersecurity and Third-Party Risk: Third Party Threat Hunting. Hoboken, NJ: John Wiley & Sons.

---

[175]   For compliance with the EU's General Data Protection Regulation, see European Union. General Data Protection Regulation Compliance Guidelines. https://gdpr.eu/.

[176]   For compliance with California's Consumer Privacy Act, see State of California, Office of the Attorney General. California Consumer Privacy Act. https://oag.ca.gov/privacy/ccpa. In July 2021, when Audi and Volkswagen suffered a data breach after a vendor left unsecured data on the internet, a California consumer sued both companies for not safeguarding personal information.

Protection and Electronic Documents Act,[177] New York's Shield Act,[178] or Singapore's Personal Data Protection Act,[179] have significantly increased the reputational and regulatory impact of inadequate noncompliant TPRM. This prospect is of particular relevance for financial services, consumer industries, health care, and government agencies, where third-party data breaches can be massive in scope and erode people's trust in the system. In response, financial service regulators have issued new frameworks and guidance documents on outsourcing, operational resilience, and third-party relationships, which stipulate fines for noncompliance.[180]

Another risk dimension is the renewed regulatory focus on environmental, social, and governance (ESG) issues, which is forcing organizations to consider not only their own environmental footprint, but also conduct appropriate risk-based assessments of their third parties and suppliers' social impact. In 2021, the European Parliament endorsed the adoption of binding EU law requiring companies to conduct environmental and human rights due diligence along their entire supply chain or face substantial fines, sanctions, and/or civil liability. In Germany, the Act on Corporate Due Diligence in Supply Chains, effective in 2023, compels companies to identify risks of human rights violations and environmental destruction at direct suppliers and, if they obtain evidence of a potential abuse, also at indirect suppliers. Organizations need to take countermeasures and document them to federal authorities or risk penalties if companies violate their due diligence obligations.[181] These new laws will likely have global implications, since companies headquartered outside the EU but with operations and employees within the EU are included in its scope.

How can TPRM programs comply not only with the flurry of new regulations but also cope better with emerging risks and strengthen their supply chains? With the surge in cyber attacks during COVID-19, many contingency plans proved inadequate to deal with the growing threat landscape. Organizations are starting to see TPRM as an essential part of business resilience. Fulfilling these requirements calls for a comprehensive solution and significant operational changes.[182] TPRM programs need to move beyond their departmental silos such as procurement, compliance, IT security, risk, and data privacy, to a cross-functional approach of monitoring and managing third-party relationships.[183] Programs are now expected to monitor multiple risk domains, including cybersecurity, data privacy, anti-bribery and anticorruption, ESG, and more. Programs also need to extend deeper into supply chains to address these risks, paying closer attention to fourth parties and beyond. Failure to do so will pose a strategic risk to organizations' ability to achieve their objectives. Interestingly, digital technology can offer solutions to manage and analyze vast amounts of data and provide useful insights on risks in the supply chain.

[177] For compliance with Canada's Personal Information Protection and Electronic Documents Act (PIPEDA), see Office of the Privacy Commissioner of Canada. PIPEDA Compliance Help. https://www.priv.gc.ca/en/privacy-topics/privacy-laws-in-canada/the-personal-information-protection-and-electronic-documents-act-pipeda/pipeda-compliance-help/.

[178] For compliance with New York's Stop Hacks and Improve Electronic Data Security (Shields) Act, see E. Kost. 2022. Meeting the 3rd-Party Risk Requirements of The NY SHIELD Act. UpGuard. 22 August. https://www.upguard.com/blog/tprm-and-the-ny-shield-act.

[179] For compliance with Singapore's Personal Data Act, see Startup Decisions. Complying with the Personal Data Act of Singapore. https://www.startupdecisions.com.sg/singapore/business-laws/personal-data-act-of-singapore-compliance/.

[180] Examples include the European Central Bank's Cyber Resilience Oversight Expectations for Financial Market Infrastructure, the Bulletin by the US Office of Comptroller of the Currency on Sound Practices to Strengthen Operational Resilience, or the Statement by the Bank of England on Outsourcing and Third-party Risk Management.

[181] Business & Human Rights Resource Centre. German Parliament Passes Mandatory Human Rights Due Diligence Law. https://www.business-humanrights.org/en/latest-news/german-due-diligence-law/.

[182] L. Kleinman. 2020. The Rise of Third-Party Digital Risk. Forbes. 14 July. https://www.forbes.com/sites/forbestechcouncil/2020/07/14/the-rise-of-third-party-digital-risk/.

[183] According to Gartner's 2019 TPRM survey, in more than of organizations, compliance (34%) and legal (25%) are the primary owners of third-party risk management, but many other functions, such as procurement, audit, and the risk management office have a stake as well.

## 5.2. Gaps and Vulnerabilities in the Third-Party Life Cycle

In 2021, the Ponemon Institute conducted a cross-industry survey of US-based TPRM professionals for an in-depth look at how organizations approach this risk at each stage of the third-party life cycle (Figure 16).[184]

The first stage of the third-party life cycle is "Source and Select," when factors such as cost, return on investment, efficiency, and time savings come into play in selecting a third party. One factor that is often overlooked is network security—the one factor that has the potential to take down an entire company. Over half of respondents (51%) stated that their organizations are not assessing the security and privacy practices of third parties before granting them access to sensitive and confidential information, instead relying on signed contracts and reputation.

At the "Intake and Score" stage, risk assessment and scoring are meant to determine the safety of onboarding a new third-party vendor. However, most organizations do not evaluate risk before a third party is engaged nor do they define or rank levels of risk—necessary steps to set the levels of security needed to defend against a breach or hacking incident. Based on the abovementioned report, 65% of third parties are not required to fill out a security questionnaire, and 74% are never asked to conduct remote or on-site assessments. The numbers are likely higher in economies that require less reporting.

During the "Identity and Access Management" stage, an organization identifies the access requirements for the recently hired third party to provide the services it was hired for. However, two- thirds of respondents report that their organizations lack a comprehensive inventory of all third parties and lack visibility into the level of access and permission for internal or external users. Among sectors, only the public sector had more than half of respondents (53%) state that they had an inventory of third parties with access to critical systems.

**Figure 16: Third-Party Life Cycle**

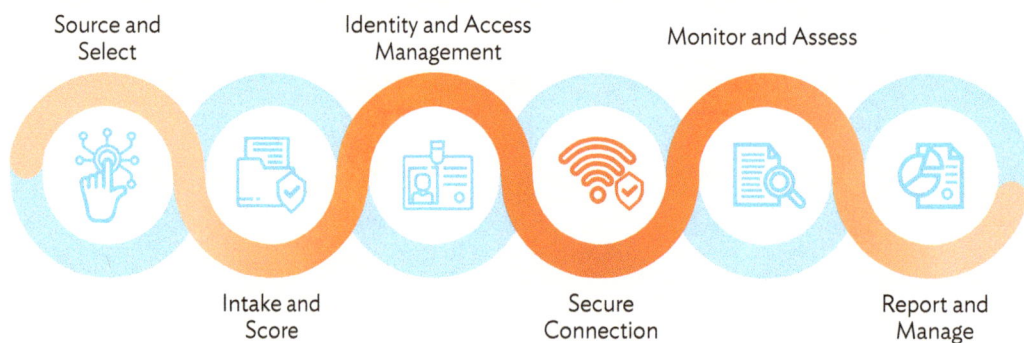

Source: Ponemon Institute and SecureLink. 2021. *A Crisis in Third-Party Remote Access Security*. Austin, Texas: SecureLink, Inc.

---

[184] The Ponemon Report is based on a survey of 627 individuals who are involved in their organization's TPRM efforts. The average number of third parties for organizations participating in this survey is 2,368. Sector representation includes financial services (113 respondents), health and pharmaceutical (69), public sector (63), services (64), industrial and manufacturing (56), retail (56), and tech and software (56). Ponemon Institute and SecureLink. 2021. *A Crisis in Third-Party Remote Access Security*. Austin, Texas: SecureLink, Inc. https://cyberdistribution.co.uk/wp-content/uploads/2021/08/SL-Report-ThirdPartySecurity.pdf.

Establishing "Secure Connections" into the organization's network and information systems is the riskiest point in the third-party life cycle, where cyber attacks often occur. Over half of respondents have experienced a data breach caused by third parties that resulted in misuse of sensitive or confidential information, with 74% of respondents stating that it was caused by granting too much privileged access. To secure their networks, organizations need to assess and evaluate their security and access protocols and have full visibility into the level of access and permissions for both internal and external users. Yet, managing third-party permissions and remote access was overwhelming for most respondents (73%), a drain on internal resources, and often subject to ambiguous roles and responsibilities between technical staff and management. As a result, organizations are most vulnerable in the connectivity stage due to a lack of control, visibility, and compliance of their third parties.

Preventative actions against potential attacks such as granting least privileged access via granular controls or preventing third parties from sharing usernames and passwords as part of a zero trust architecture tend to be largely ignored. Not surprisingly, two-thirds of respondents lack confidence that third parties would report a data breach involving confidential information. Without proof of third party's involvement, organizations cannot enforce accountability and bear the liability for the breach. At this stage, proper auditing capabilities are critical to log the activity of each third party during network sessions and trace any suspicious activity to its source.

Once a third party's network connectivity has been established, the activity should be subject to continuous "Monitoring and Assessment." Yet, most organizations (54%) are failing to keep track of the security and privacy practices of third parties on an ongoing basis. Instead of using automated monitoring tools, organizations tend to rely on periodic legal or procurement reviews or trust the reputation of the third party.

The last stage of the life cycle involves "Reporting and Management." Most respondents (59%) state that there is no centralized oversight over third parties. Only a third of respondents are submitting regular reports to their board of directors on the effectiveness of the TPRM and potential risks to the organization. Without centralized control and clearly defined management responsibilities, reporting and managing these complex relationships remains an unresolved challenge.

To address the disconnect between an organization's perceived third-party access threat and the security measures it employs, the solution requires eliminating many of these common oversights and poor third-party practices. According to industry observers, organizations are starting to adapt their TPRM programs, but more needs to be done.[185] In particular, organizations underestimate the full complexity of a comprehensive TPRM solution and tend to focus narrowly on individual components and contend with insufficient budgets.

Digital solutions to automate TPRM tasks are expected to mature within years and are already freeing up resources to focus on critical risks that require human review.

---

[185] KPMG. Third-Party Risk Management Outlook 2022. https://home.kpmg/xx/en/home/insights/2022/01/third-party-risk-management-outlook-2022.html.; and Prevalent. The 2022 Third-Party Risk Management Study. https://www.prevalent.net/content-library/2022-third-party-risk-management-study/?aliId=eyJpIjoiOEVlS3NaTDhiOUhTbW1sVyIsInQiOiJyd3lPNkZCUjRMbWJXaW11QIdqV0RnPT0ifQ%253D%253D.

## 5.3. Framework and Processes for Third-Party Risk Management Integration

According to Gartner, more than 80% of executives acknowledged that third-party risks were identified after initial onboarding and due diligence, highlighting the shortcomings of traditional methods.[186] The rapidly changing business environment demands new approaches that account for the increase in outsourcing and collaboration. As these partnerships become increasingly complex, senior management and boards have to rely on various solutions and methods to evaluate third parties. An effective TPRM framework and process that can be integrated with the organization's overall risk management should include the following steps:

**Step 1: Strategic Analysis.** Strategic risk reduction is the main goal of TPRM programs. While programs still remain primarily focused on tracking information security risks, followed by data privacy and protection—typically the domains of IT security teams—organizations are paying more attention to non-IT risks, notably compliance and ethics, reputation (sanctions, adverse media coverage) and finance, as well as business continuity (geopolitical unrest). By unifying non-IT risk intelligence with the results of traditional cybersecurity and data privacy assessment, supplier risks gain relevance among multiple departments and elevate the strategic value of TPRM.

The assessment phase plays a pivotal role prior to onboarding a third party to identify the various risks that could be introduced to the organization. Given the growth in third-party networks, an increasingly popular approach is to use security ratings to determine whether the external security posture of third (and fourth) parties meets a minimum accepted score. Traditional risk assessment methodologies like penetration testing, security questionnaires, and on-site visits, are time-consuming, expensive, and often rely on subjective assessments and vendors' claims about their information security controls. By using security ratings, which are compiled by an independent security rating platform, in conjunction with existing risk management techniques, third-party risk management teams can have a data-driven, objective, and dynamic measurement of an organization's security posture. These risk ratings could become as important as credit ratings by providing a real-time picture of risks posed by supply chains and enabling governments to better assess and manage their vendors' cybersecurity performance throughout the life cycle.

**Step 2: Engagement.** If the vendor's security rating is considered sufficient, the next step is for the vendor to complete a questionnaire that provides insights into their security controls not visible to outsiders. Manual methods still persist despite the dissatisfaction with time-consuming spreadsheet questionnaire filling. More recently, dedicated solutions with questionnaire libraries offer help with identifying potential weaknesses among third-party vendors and business partners that could result in a data breach. Consistency in assessing all vendors against the same standardized checks is key to ensure comparability. Many organizations make the mistake of believing that they do not need to monitor low-risk third parties, such as marketing tools or office cleaning services.

---

[186] K. Lewis and K. Clinton. Panel: The Key to Third-Party Risk Management in APAC-Aligned Assurance. Gartner (webinar). On demand. https://www.gartner.com/en/webinars/4014198/panel-the-key-to-third-party-risk-management-in-apac-aligned-assurance.

**Step 3: Remediation.** Organizations need to establish policies and processes for evaluating and vetting third parties' security practices based on an acceptable level of third-party risk. This includes building a comprehensive inventory of all third-party relationships, knowing the sensitivity of the data and how it is accessed and stored, and having clarity which third parties have access to what type of data. Many organizations fail to provide context around these assessments, even though different types of vendor relationships can pose different levels of risks. For instance, a supplier may only be used to transfer nonsensitive information such as a blog post, while another supplier may handle, store, and process sensitive data of customers. If the third party has unacceptable risks, the organization may decide not to collaborate with them until they fix the security issues.

The challenge for organizations is to keep track of hundreds or even thousands of third parties and monitor their security status on an ongoing basis. A remediation approach should focus on the most critical priorities first to ensure that the risks are resolved quickly and with an audit trail. According to industry surveys, it takes about 2.5 weeks from the time of learning of an incident to receiving a confirmation of remediation—a lifetime for an organization that is vulnerable to a potential data exploit.[187] Automating this type of incident response would reduce costs and time by centrally managing all vendors in a single platform, identifying at-risk parties, and quickly mitigating risks based on prescriptive remediation guidance. A platform with remediation workflows would allow an organization to request remediation from a specific vendor based on automated scanning and completed questionnaires. In addition, penetration testing (or ethical hacking) could be used to probe exploitable security vulnerabilities.

**Step 4: Approval.** After remediation (or lack thereof), the organization can decide whether to onboard the vendor or choose to look for a different vendor based on risk tolerance, the criticality of the vendor, and any compliance requirements that need to be met.

A framework for third-party approvals can be highly complex due to the distributed risk decision-making environment within organizations (Figure 17) and an "overabundance of ownership" (Table 13). Stakeholders, including board members, business leaders, and employees, complain about contradictory narratives produced by different functions, auditing fatigue, and lack of clarity over decision rights and follow-up. Given heightened stakeholder oversight of TPRM initiatives, organizations are pushing for increased accountability and a coordinated, consolidated view into TPRM activities. However, organizations struggle to establish TPRM governance as management functions are divided between due diligence and continuous monitoring.

One way of addressing these tensions and strengthening risk management activities is to implement TPRM activity governance and provide 360-degree visibility and assurance to the board, regulators, and customers. A risk committee would bring the different third-party risk owners and subject matter experts together to analyze current risks, plan for emerging ones, and evaluate control effectiveness. Aligning assurance processes can be achieved by building an assurance map, adopting a collaborative risk assessment, coordinating reporting efforts and metrics, and investing in continuous learning and maturity efforts, thereby promoting unified decision-making and eventually improving risk management outcomes. After establishing

---

[187]   Prevalent. 2022 Third-Party Risk Management Study.

**Figure 17: The Distributed Risk Decision-Making Environment**

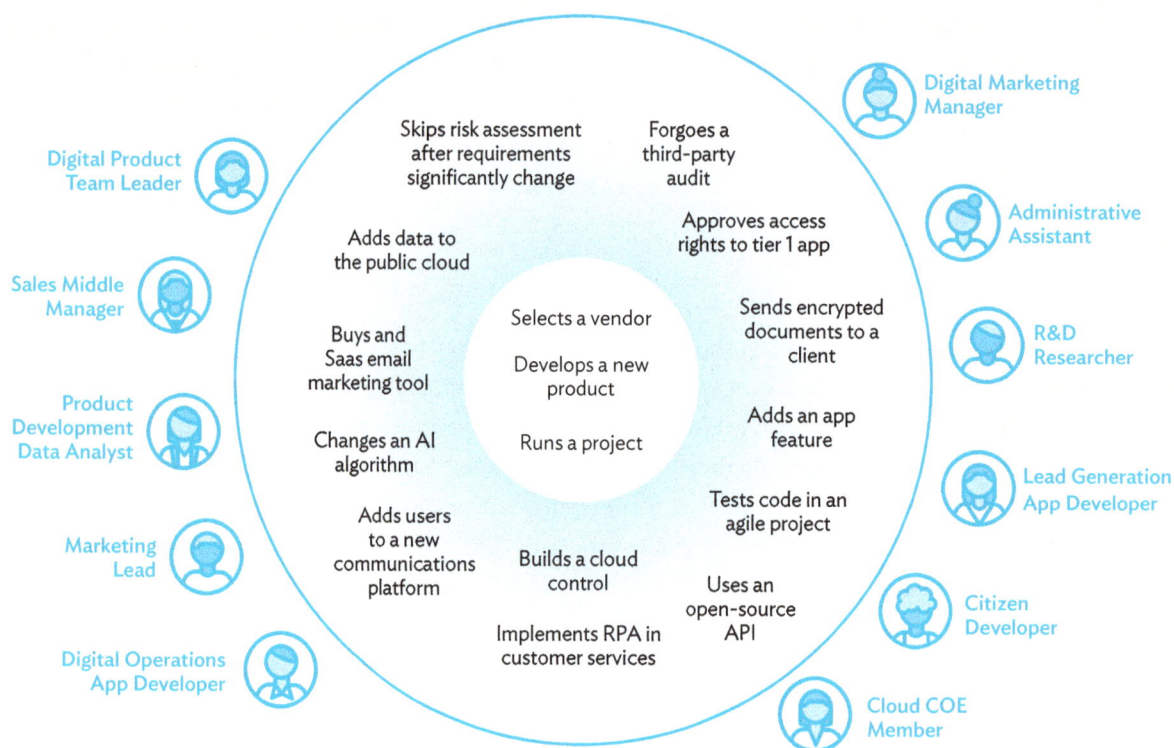

AI = artificial intelligence, API = application programming interface, COE = center of excellence, RPA = robotic process automation, R&D = research and development, SaaS = software as a service.
Source: D. Hebda. Maximize Third-Party Risk Management with Aligned Assurance. *Gartner* (webinar). On-demand.

**Table 13: Who is Responsible for Third-Party Risk Management Activities?**
(%)

| | | Audit | Compliance | Finance | IT/IS | Legal | Privacy | Procurement | Quality | Risk (ERM) | TPRM Office | Others |
|---|---|---|---|---|---|---|---|---|---|---|---|---|
| **Due diligence** | Prospecting | 22 | 27 | 16 | 19 | 21 | 13 | 33 | 14 | 23 | 19 | 2 |
| | Complete business justification form | 2 | 14 | 27 | 17 | 18 | 24 | 9 | 24 | 12 | 18 | 15 |
| | Send and collect third-party questionnaire | 5 | 3 | 15 | 28 | 14 | 16 | 22 | 12 | 25 | 13 | 19 |
| | Send and collect third-party codes of conduct, policies, etc. | 18 | 3 | 2 | 13 | 29 | 15 | 20 | 24 | 13 | 25 | 11 |
| | Remediate risk | 17 | 17 | 2 | 3 | 16 | 33 | 18 | 19 | 29 | 14 | 22 |
| | Contracting | 16 | 28 | 23 | 4 | 2 | 11 | 22 | 17 | 16 | 36 | 14 |
| **Ongoing management** | Monitoring/auditing | 26 | 13 | 20 | 16 | 2 | 1 | 33 | 30 | 13 | 13 | 18 |
| | Recertification | 12 | 18 | 13 | 18 | 17 | 3 | 2 | 17 | 31 | 15 | 11 |

ERM = ERM = Enterprise Risk Management, IS = Information Security, IT = Information Technology, TPRM = Third-Party Risk Management.
Note: N = 953, 2019 State of the legal compliance. Audit and risk functional benchmarking.
Source: *Gartner*.

a risk committee, organizations should identify a single, primary owner responsible for TPRM who identifies and mitigates the most critical third-party risks for the organization and requires unrestricted access to critical third-party risk information.

**Step 5: Monitoring and Reporting.** It is important to continue the monitoring of third parties once they have been onboarded, are able to access the organization's information systems and sensitive data, and have become familiar with internal business processes. Automation tools are key to ensuring that only authorized users have access to sensitive data, that fraud attempts by third parties can be detected and responded to in real time, and that follow-up risk assessments are triggered automatically when thresholds are exceeded. Continuous security monitoring is a threat intelligence approach that automates the monitoring of information security controls, vulnerabilities, and other cyber threats to support risk management decisions.[188]

Reporting on the results of a TPRM program is becoming more complex and time-consuming, but results show that these programs are effective. In the last 4 years, legal and compliance leaders have classified 2.5 times more third parties as high-risk. A 2022 industry survey of third-party risk audits showed that 74% of respondents had to report on third-party data privacy and protection controls, followed by information security controls (57%), ESG topics (23%), and human trafficking and slavery regulations (18%) (footnote 241). These audits can occur in yearly intervals (42%), quarterly (21%), or on an ad hoc basis (23%), and take anywhere between 1 week, 1 month, or longer to produce the evidence for a regulatory audit.

Organizations can rely on several standards to benchmark their TPRM programs. Both the National Institute of Standards and Technology (NIST) and the International Organization for Standardization (ISO) have developed widely used risk management frameworks that offer useful guidance for assessing any TPRM program.[189]

---

[188]   M. Tarchinski Krzoska. 2022. Making the Business Case for Continuous Monitoring. *Audit Board* (blog). 21 February. https://www.auditboard.com/blog/making-the-business-case-continuous-monitoring/.

[189]   J. Boyens et al. 2021. Key Practices in Cyber Supply Chain Risk Management: Observations from Industry. *NIST Interagency or Internal Report*. No. 8276. Gaithersburg, MD: National Institute of Standards and Technology. https://doi.org/10.6028/NIST.IR.8276.

# 6

# PRIVACY RISK AND DATA PROTECTION: BUILDING TRUST THROUGH GOVERNANCE

Trust is a crucial currency in today's data economy—and consumer privacy violations and data negligence are the two quickest ways to devalue it.

Over the last few years, a convergence of changing consumer expectations, digital market competition, and accelerating government action has ushered in new rules on data privacy and data sharing, which have emerged as the organizing principles of the data economy. Rather than treating data as a free resource that users readily allow to be appropriated, most countries have begun to treat personal data as an asset that belongs to individuals and is held in temporary trust by third-party entities.

This chapter will attempt to put this new dynamic into full view and highlight key frameworks, standards, and tools affecting privacy risk and data protection.

## 6.1. Trust Building

Trust needs to be created and maintained between the parties who share, collect, and use data. If data are made available to people and organizations who need it, it can enable better and faster decision-making, catalyze innovation, create more efficient and effective services and products, and fuel economic growth and productivity. Evidence supports the argument that data and better data flows create value not only in monetary and financial terms,[190] but also as a societal value for individuals and communities.[191]

Countries cannot fully reap the benefits that digital tools and data integration offer without creating and maintaining public trust. To build and maintain this trust, governments are playing a key role, be it by setting clear and enforceable rules to protect citizens' rights, embedding accountability and transparency in public data systems, or taking steps to curtail the power of dominant tech companies. These rules need to reflect and fit the country context —with the caveat that some governments may seek to forego the need to build trust by using data and digital tools to exert more control.[192] While different ecosystems are unique, efforts should be made to build trust in data sharing horizontally, across all sectors, rather than purely on a sector-by-sector basis.

---

[190] Studies suggest that data sharing can help generate social and economic benefits worth between 0.1% and 1.5% of gross domestic product (GDP) in the case of public sector data, and between 1% and 2.5% of GDP when public sector and private sector data are combined. If scaled up to the GDP of the 20 largest economies in 2019, data sharing could unlock between $700 billion and $1.75 trillion in value. A 25% increase in data sharing could generate an additional $47 billion to $118 billion worldwide. Open Data Institute. The Economic Impact of Trust in Data Ecosystems–Frontier Economics for the ODI. https://theodi.org/article/the-economic-impact-of-trust-in-data-ecosystems-frontier-economics-for-the-odi-report/.

[191] World Bank. World Development Report 2021: Data for Better Lives. https://www.worldbank.org/en/publication/wdr2021. (see chapters 1 to 4).

[192] Contextual factors are crucial to understanding the role of trust in influencing data sharing. Norms and unwritten attitudes will play a role in determining baseline levels of trust. These norms are likely to vary substantially both between individuals and across different regions.

The challenge facing governments is how to establish rules that protect citizens from harm while at the same time encouraging useful innovation. As of March 2022, 157 economies have enacted privacy laws.[193] Several factors have driven the rapid increase in data protection and privacy laws, including the catalytic effect of the EU GDPR, growing awareness of the risks of data misuse, the desire to create an enabling framework for responsible innovation, and the need to meet donor standards on data protection.[194]

The EU's GDPR, adopted in 2018, has altered the global data protection landscape by providing a more rigorous model for protecting the privacy of individual data through mechanisms that strengthen individual control over how data is used and make data controllers more accountable. By introducing the concept of extraterritorial applicability, the GDPR has prompted a global expansion of data protection frameworks. The GPDR also explicitly incorporates the *Privacy by Design Principles*, which seek to ensure that personal data are protected by default in any given IT system or business practice.

Although the GDPR serves as a global template of sorts for most of these laws, many economies have adopted different implementation modalities. Some GDPR principles, like data breach notification, data portability, and demonstrable accountability, have become commonplace. Most new laws also include adequacy-like data export controls based on the extent to which the receiving economy's laws approximate their own (Sri Lanka, the PRC) and provide contractual protections. While all laws introduce specialized enforcement bodies, most fail to create genuinely independent data protection agencies (DPAs).[195] As governments and organizations continue to demand protection for data transferred outside their national borders, more jurisdictions are putting in place data localization requirements, even though this adds significant costs to data operations without necessarily improving security.[196]

Privacy legislation continues to be very well received around the world. According to CISCO's 2022 Data Privacy Benchmarking Study, 83% of all corporate respondents stated that privacy laws have had a positive impact.[197] Many geographies, including Indonesia, the Philippines, the PRC, Thailand, and Viet Nam, had 90% or more of respondents stating that privacy regulation has had a positive impact (Figure 18).

---

[193] Some of the most important data privacy legislation enacted in 2021 includes the PRC, which revised its existing patchwork of laws into a comprehensive Personal Information Privacy Law. In addition, India and Indonesia are preparing comprehensive data protection bills to replace their existing noncomprehensive laws. Other economies enacted significant updates to existing laws in 2021–2022, including Viet Nam; the Republic of Korea; Hong Kong, China; Kazakhstan; Uzbekistan; Sri Lanka; and Mongolia. Bangladesh, Pakistan, and Brunei Darussalam have official bills or draft laws at various stages of the political or legislative process. G. Greenleaf. 2022. Now 157 Countries: Twelve Data Privacy Laws in 2021/22. 176 *Privacy Laws & Business International Report*. 1, 3–8. https://papers.ssrn.com/abstract=4137418.

[194] For instance, the World Bank is developing a set of minimum data privacy standards that it will eventually require economies to adopt or maintain before funding digital- and data-related projects. Humanitarian aid organizations have long recognized the importance of data privacy and ethical data use in their work with vulnerable groups.

[195] The failure of such a high percentage of laws to create independent DPAs presents problems for those economies seeking to (i) obtain a positive adequacy assessment from the EU, (ii) accede to the Council of Europe's Convention 108+, and (iii) join the Global Privacy Assembly as full members. C. de Terwangne, 2021. Council of Europe Convention 108+: A Modernized International Treaty for the Protection of Personal Data. Computer Law & Security Review. 40. https://doi.org/10.1016/j.clsr.2020.105497.; and Global Privacy Assembly. https://globalprivacyassembly.org/.

[196] See also ADB and Amazon Web Services Institute. 2022. Data Management Policies and Practices in Government. Manila. https://www.adb.org/publications/data-management-policies-strategies-government.

[197] Cisco. 2022. *Data Privacy Benchmark Study*. San Jose.

## Figure 18: The Impact of Privacy Laws
### (%)

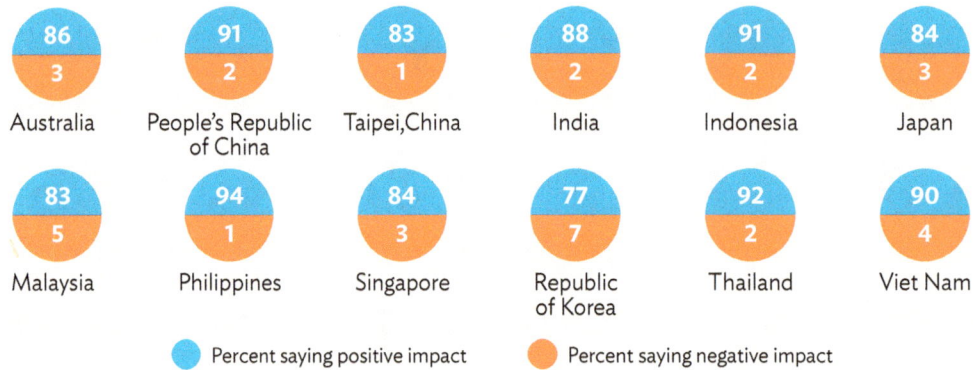

| | | | | | |
|---|---|---|---|---|---|
| **86** / 3 | **91** / 2 | **83** / 1 | **88** / 2 | **91** / 2 | **84** / 3 |
| Australia | People's Republic of China | Taipei,China | India | Indonesia | Japan |
| **83** / 5 | **94** / 1 | **84** / 3 | **77** / 7 | **92** / 2 | **90** / 4 |
| Malaysia | Philippines | Singapore | Republic of Korea | Thailand | Viet Nam |

● Percent saying positive impact  ● Percent saying negative impact

Source: Cisco. Data Privacy Benchmark. 2022.

For companies, data privacy has become a top-level business imperative.[198] In CISCO's global privacy survey, 90% of respondents indicated they would not buy from an organization that does not properly protect its data, while 91% consider external privacy certification important in their buying decisions.[199] With three out of four global consumers claiming to pay close attention how companies collect and use their data, trust and transparency in an organization are the most important factors driving consumer willingness to share personal information.

At the same time, nearly half of consumers (46%) feel they do not fully understand what organizations are doing with their data. This sentiment is borne out by the fact that 90% of companies acknowledge that they have amassed massive data silos, which are expensive to maintain and operationalize.[200] An internal Facebook document revealed that the company has no idea where all of its first-party, third-party, and sensitive-categories data goes or how it is processed.[201] Likewise, Google Analytics is facing considerable challenges with privacy issues.[202] Shortcomings in handling data create valid grounds for regulatory investigations and foreshadow the need for future changes in the curation and processing of data.

Consumer preferences offer new opportunities for market differentiation. This became evident in the wake of Apple's release of its iPhone 14.5 operating system in 2021. The new operating system gave users the option to shut down third-party tracking across their iPhone apps—at a cost of $10 billion in lost revenues

---

[198] Privacy metrics are routinely reported to boards of directors. J. Polonetsky and O. Tene. 2022. Measuring Privacy Programs: The Role of Metrics. *International Association of Privacy Professionals*. 24 March. https://iapp.org/news/a/measuring-privacy-programs-the-role-of-metrics/.

[199] Survey included 5,000+ security professionals from 27 countries (including Australia, India, Indonesia, Japan, Malaysia, the Philippines, the PRC, Singapore, the Republic of Korea, Thailand, and Viet Nam). Cisco. 2022. *Data Privacy Benchmark Study*. San Jose.

[200] MuleSoft. 2021. Study Reveals Integration Challenges Threaten Digital Transformation, With Organizations Spending on Average US$3.5 Million on Custom Integration Labor Costs. https://www.mulesoft.com/press-center/march-2021-connectivity-benchmark-report.

[201] L. Franceschi-Bicchierai. 2022. Facebook Doesn't Know What It Does With Your Data, Or Where It Goes: Leaked Document. *Vice* (blog). 26 April. https://www.vice.com/en/article/akvmke/facebook-doesnt-know-what-it-does-with-your-data-or-where-it-goes.

[202] Erin. 2022. Google Analytics Privacy Issues: Is It Really That Bad? *Matomo* (blog). 2 June. https://matomo.org/blog/2022/06/google-analytics-privacy-issues/.

for major data harvesters and social media sites in the second half of 2021.[203] Meta, Facebook's parent company, expects that being cut off from this data trail will amount to another $10 billion in lost revenue in 2022 alone. For Apple, this step solidified the brand's reputation for privacy protection and allowed it to draw in new privacy-conscious users.

Most global consumers (82%) see data exchange as essential for the running of a modern society and are prepared to engage with the data economy, according to the Global Data Privacy Report 2022 by the Global Data & Marketing Alliance.[204] Across 16 global markets, 53% of consumers describe themselves as "Data Pragmatists,"[205] people who are ready to exchange data with businesses so long as they perceive a clear benefit for doing so; another 29% can be described as "Data Unconcerned," people who show little or no concern about their data privacy. Less than one in five (18%) falls into the category of "Data Fundamentalists," people who are unwilling to share personal information (Figure 19).

**Figure 19: Attitudes toward Privacy in 16 Global Markets (%)**

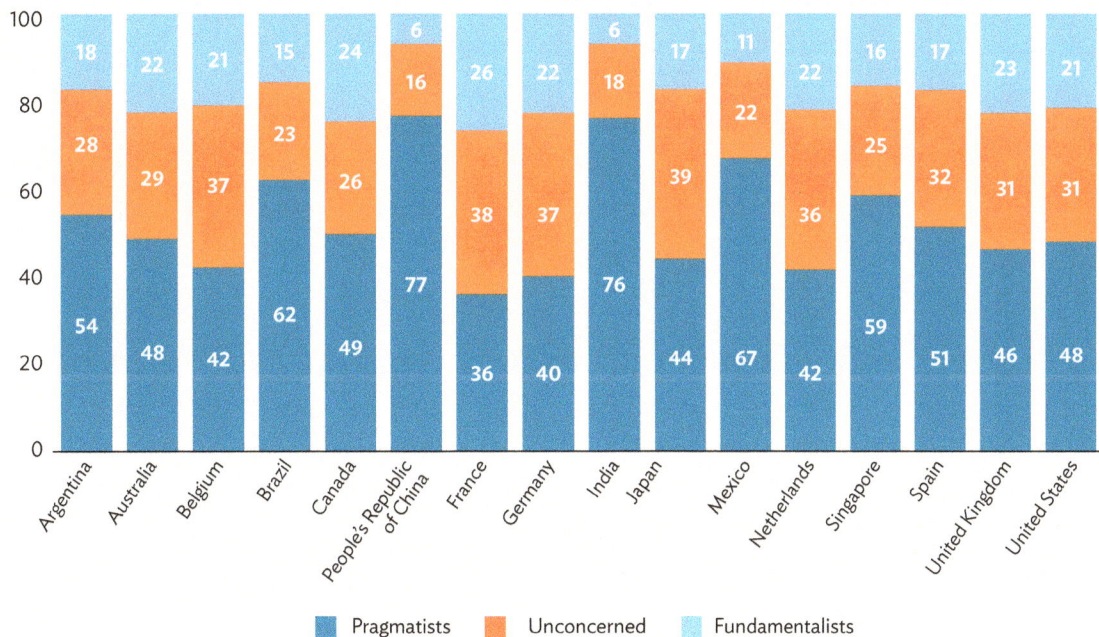

Source: GDMA. Consumer Attitudes: Global Data Privacy Report.

---

203 The $300 billion mobile advertising industry relies on vast amounts of consumer data to deliver targeted ads. After Apple's App Tracking Transparency announcement, the opt-out rate reached close to 90%; Facebook's opt-out rate in the US topped 96%, reflecting its poor handling of customer data. Target Internet. How Apple IOS14 Changes Have Affected Facebook Third-Party Tracking. https://www.targetinternet.com/resources/how-apple-ios14-changes-have-affected-facebook-third-party-tracking/.; and K. Jerath. 2022. Mobile Advertising and the Impact of Apple's App Tracking Transparency Policy.

204 In total, 20,626 respondents aged 18+ were surveyed across the 16 markets in December 2021. The survey was conducted by Foresight Factory on behalf of Acxiom, a customer intelligence company; and GDMA, an organization representing marketing associations. Global Data and Marketing Alliance. 2022. Global Data Privacy: What the Consumer Really Thinks. https://globaldma.com/consumer-attitudes/.

205 It is worth noting that the share of data pragmatists varies considerably across countries. For example, in India and the PRC, over three in four consumers fall within the pragmatist mindset.

With growing public awareness of data protection regimes, consumers increasingly believe they can exercise greater control over the exchange of their personal data. Global consumers show high levels of responsibility for their own data security, but also growing expectations of industry.[206] Almost three in four consumers across all 16 markets hold the view that their data are their property and that they should be able to trade it if they wish.

These three factors—accelerating government action, intense market competition, and consumer mistrust—are converging toward a point where individuals will be able to exercise greater control over their personal data. While consumers still seek the convenience and benefits from their data, they will be keen to set the terms over what data they share and who they share it with.

## 6.2. Managing Privacy Risk

Managing privacy requires managing privacy risk. Privacy and data protection programs are being challenged by a proliferation of new regulatory requirements. The latest examples involve California's Age-Appropriate Design Code Act to force tech companies to take more responsibility to protect children online[207] and the EU's Digital Markets Act, which could bring major changes to what people can do with their devices and apps and prevent dominant platforms from preferencing their own apps and services.[208]

Prompted by such regulatory changes, organizations will need to evolve their privacy programs continuously to meet new requirements. This assessment is also echoed by a global survey of risk and compliance practitioners, who ranked regulatory compliance (66%) and privacy, and data protection and security (64%) as top priorities for 2022, well ahead of more familiar topics such as anticorruption and fraud, ESG, organizational culture, and diversity and inclusion.[209]

How can privacy and data protection leaders adapt to continuous regulatory changes impacting their organization? To address these issues, there first must be clarity on how to calibrate the organization's appetite for regulatory risk; understand the relationship between privacy, cybersecurity, and organizational risk and how these are managed; secure demonstrable buy-in from senior management; and, lastly, decide on the governance model and metrics being used (Figure 20).

The purpose of privacy frameworks is to help organizations deal with regulatory changes. They provide a clear structure and help shape key policies and procedures to support the privacy program and its stakeholders. Upon selecting a privacy framework, one of the first steps is to map out key control areas (to identify overlaps between privacy and security controls) and define data validation maps (for the processing of personal and sensitive data) before assessing how this framework aligns with

---

[206] For instance, the share of consumers in the PRC (31%), India (26%), and Japan (22%) that claim brands should hold final responsibility for data security is significantly higher than the global average of 14%.

[207] P. Ceres. 2022. The US May Soon Learn What a "Kid-Friendly" Internet Looks Like. *Wired*. 1 September.

[208] K. Johnson. 2022. Europe Prepares to Rewrite the Rules of the Internet. *Wired*. 28 October. https://www.wired.com/story/europe-dma-prepares-to-rewrite-the-rules-of-the-internet/.

[209] NAVEX. 2022 Definitive Risk and Compliance Benchmark Report. https://cdn.navex.com/image/upload/v1662757639/resource%20documents/2022_Definitive_R-C_Benchmark_Rpt_INT.pdf?_gl=1*wfb425*_ga*NDYxNDg3NjczLjE2NjY4MzcONjY.*_ga_JRYF9MG532*MTY2NjgzNzQ2Ni4xLjEuMTY2NjgzNzUwMS4wLjAuMA.

**Figure 20:** The Relationship between Privacy, Cybersecurity, and Organizational Risk

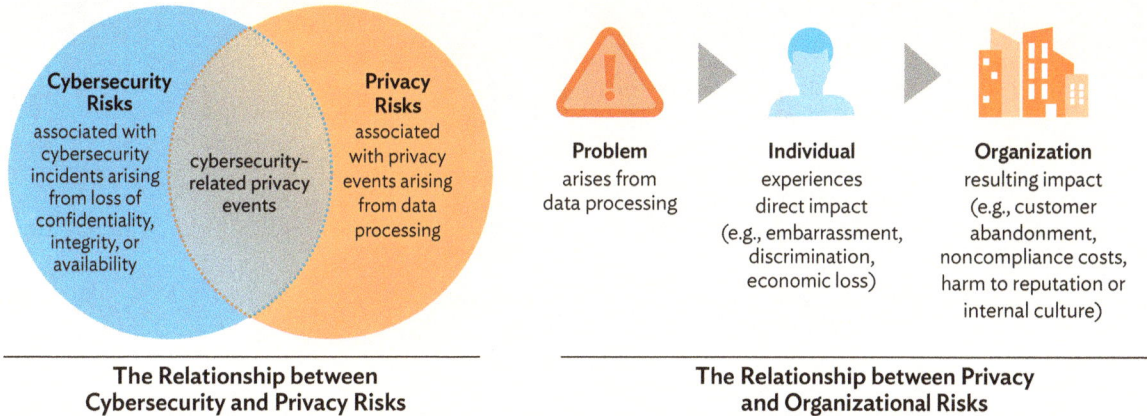

The Relationship between Cybersecurity and Privacy Risks

The Relationship between Privacy and Organizational Risks

Source: NIST. 2020. *NIST Privacy Framework: A Tool for Improving Privacy Through Enterprise Risk Management, Version 1.0.* Gaithersburg, Maryland: National Institute of Standards and Technology.

regulations the organization must comply with. It may be necessary to tailor the chosen framework to specific privacy risks and regulatory requirements. Once the framework is up and running, it can be leveraged every time a regulatory change is forthcoming by mapping the new requirements into the controls that have been documented and adjusting as necessary. The privacy office was most often located in IT (37%), followed by Security (34%), Compliance (11%), Legal (9%), and Operations (8%). Several privacy frameworks are available, including the NIST Privacy Framework (Table 14),[210] the ISO 27701 Privacy Information Management Standard,[211] or OECD's Privacy Framework.[212]

**Table 14:** Summary of the National Institute of Standards and Technology Privacy Framework

| Functions | Categories (Groups of privacy outcomes tied to programmatic needs and particular activities) | Privacy Management Activities | Responsible Business Group |
|---|---|---|---|
| **Identify**<br><br>Develop the organizational understanding to manage privacy risk for individuals arising from data processing | • **Inventory and mapping.** Data processing by systems, products, or services is understood and informs the management of privacy risk.<br>• **Business environment.** Organization's mission, objectives, stakeholders, and activities are understood and prioritized. This information is used to inform privacy roles, responsibilities, and risk management decisions.<br>• **Risk assessment.** The organization understands the privacy risks to individuals and how such privacy risks may create follow-on impacts for operations, including mission, functions, workforce, reputation, etc. | • Deliver self-assessment/privacy risk assessment reports<br>• Create a data inventory/map of personal information | IT, Legal, Compliance |

*continued on next page*

---

210   NIST. 2020. *NIST Privacy Framework: A Tool for Improving Privacy Through Enterprise Risk Management, Version 1.0.* Gaithersburg, Maryland: National Institute of Standards and Technology. https://www.nist.gov/privacy-framework/privacy-framework.

211   ISMS Online. ISO 27701–The Standard for Privacy Information Management. https://www.isms.online/iso-27701/.

212   OECD. Privacy. https://www.oecd.org/sti/ieconomy/privacy.htm.

**Table 14** *continued*

| Functions | Categories (Groups of privacy outcomes tied to programmatic needs and particular activities) | Privacy Management Activities | Responsible Business Group |
|---|---|---|---|
| | • **Data processing ecosystem risk management.** The organization's priorities, constraints, and risk tolerance are established and used to support risk decisions associated with managing privacy risk and third parties within the data processing ecosystem. The organization has established and implemented the processes to identify, assess, and manage privacy risks within this ecosystem. | | |
| **Govern**<br><br>Develop and implement the organizational governance structure to enable an ongoing understanding of the organization's risk management priorities that are informed by privacy risk | • **Governance policies, processes and procedures** to manage and monitor the organization's regulatory, legal, risk, environmental, and operational requirements and inform the management of privacy risk.<br>• **Risk management strategy.** The organization's priorities, constraints, risk tolerances, and assumptions are established and used to support operational risk decisions.<br>• **Awareness and training.** Workforce and third parties engaged in data processing are provided privacy awareness education and are trained to perform their privacy-related responsibilities consistent with policies, procedures, and privacy values.<br>• **Monitoring and review.** Policies, processes, and procedures for ongoing review of the organization's privacy posture are understood and inform the management of privacy risk. | • Establish privacy governance/ committee<br>• Implement vendor risk management program<br>• Regularly monitor, review, and update privacy policies and procedures | HR, Legal, Compliance, Business/ Marketing, IT, Information Security |
| **Control**<br><br>Develop and implement appropriate activities to enable organizations or individuals to manage data with sufficient granularity to manage privacy risks | • **Data processing policies, processes, and procedures** are maintained and used to manage data processing, consistent with the organization's risk strategy.<br>• **Data processing management.** Data are managed consistent with the organization's risk strategy to protect individuals' privacy, increase manageability, and enable the implementation of privacy principles (e.g., individual participation, data quality, data minimization)<br>• **Disassociated processing.** Data processing solutions increase disassociability consistent with the organization's risk strategy to protect individuals' privacy and enable implementation of privacy principles | • Develop privacy policies (e.g., internal privacy policy, consent management, data retention, privacy by design)<br>• Create data subject/ consumer rights request policy<br>• Implement data protection impact assessments/ privacy risk assessments | HR, Legal, Compliance, Business/ Marketing, IT, Information Security |
| **Communicate**<br><br>Develop and implement appropriate activities to enable organizations and individuals to have a reliable understanding and engage in a dialogue about how data are processed and associated privacy risks | • **Communication policies, process, and procedures** are maintained and used to increase transparency of the organization's data processing practices (e.g., purpose, scope, roles, and responsibilities in the data processing ecosystem; management commitment) and associated privacy risks.<br>• **Data processing awareness.** Individuals and organizations have reliable knowledge about data processing practices and privacy risks, and effective mechanisms are used and maintained to increase predictability consistent with the organization's risk strategy to protect individuals' privacy. | • Deliver privacy training and awareness<br>• Deliver training to employees handling personal information or personal data | HR, Legal, Compliance |

*continued on next page*

Table 14 *continued*

| Functions | Categories (Groups of privacy outcomes tied to programmatic needs and particular activities) | Privacy Management Activities | Responsible Business Group |
|---|---|---|---|
| **Protect**<br><br>Develop and implement appropriate data processing safeguards | • **Data protection policies, processes, and procedures** are maintained and used to manage the protection of data.<br>• **Identity management, authentication, and access control.** Access to data and devices is limited to authorized individuals, processes, and devices, and is managed consistent with the assessed risk of unauthorized access.<br>• **Data security.** Data are managed consistent with the organization's risk strategy to protect individuals' privacy and maintain data confidentiality, integrity, and availability.<br>• **Maintenance** and repairs are performed consistent with policies and processes.<br>• **Protective technology.** Security solutions are managed to ensure the security and resilience of systems, products, services, and associated data. | • Create and manage written information security policy<br>• Implement technical and organizational security measures to protect personal information or personal data<br>• Develop security policies and procedures (e.g., acceptable use, access control, incident response, business continuity, disaster recovery) | IT, Information Security |

HR = human resources, IT = information technology.
Source: NIST. 2020. *NIST Privacy Framework: A Tool for Improving Privacy Through Enterprise Risk Management, Version 1.0.* Gaithersburg, Maryland: National Institute of Standards and Technology.

NIST's Privacy Framework (summarized in Table 14), released in 2020, is a widely accepted, voluntary, technology-neutral industry standard. The framework can be used to implement privacy engineering, design products and services to protect individuals' privacy, and measure and improve an organization's privacy program. It can also serve as a competitive differentiator in relation to clients and customers. The privacy framework is flexible enough to address diverse privacy needs, enable innovative and effective solutions that can lead to better outcomes for individuals and organizations, and stay current with technology trends, such as the rapid expansion of AI and IoT solutions.

NIST defines privacy risk in reference to potential problems that may arise from data processing and their impacts for individuals (Figure 21). These include dignity-type effects (embarrassment or stigmas) as well as more direct harms, such as discrimination, economic loss, or physical harm.[213] Privacy risk can result from the loss of either confidentiality, integrity, or availability of personal data or from an insufficient level of predictability, manageability, or disassociability in relation to personal data (which are failures to meet privacy engineering objectives).[214] NIST's privacy risk model calculates risk based on the *likelihood* of a *problematic data action* multiplied by its *impact*.[215]

---

[213] P. Foitzik. 2020. How to Manage Privacy Risk under Both the GDPR and CCPA. *International Association of Privacy Professionals.* 25 February. https://iapp.org/news/a/how-to-manage-privacy-risk-under-both-the-gdpr-and-ccpa/.

[214] These privacy engineering principles are key for embedding privacy into an organization's technology stack. *Predictability* implies that users can understand how their data is processed by an information system; *Manageability* refers to users having granular control of PII, including alteration, deletion, and selective disclosure, enabled by identity and access management platforms; and *Dissociability* means that users can reveal selective information about themselves without revealing their identity. By minimizing access to PII, this is a privacy engineering layer where cryptography can be applied to obfuscate data and where techniques such as anonymity, de-identification, unlikability, or pseudonymity are of importance.

[215] *Impact* is an analysis of the costs should the problem occur. Since organizations may not experience these problems directly and individuals' experience may be subjective, impact may be difficult to assess accurately.

**Figure 21: Components of Privacy Engineering**

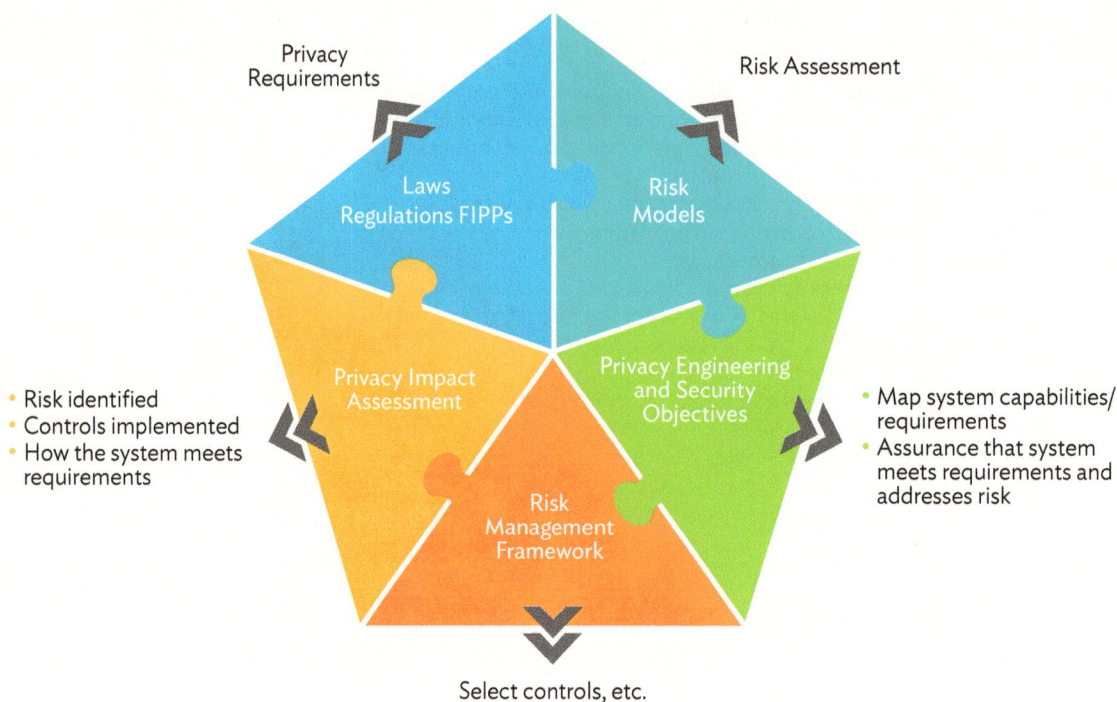

Privacy Requirements

Risk Assessment

Laws Regulations FIPPs

Risk Models

Privacy Impact Assessment

Privacy Engineering and Security Objectives

- Risk identified
- Controls implemented
- How the system meets requirements

- Map system capabilities/ requirements
- Assurance that system meets requirements and addresses risk

Risk Management Framework

Select controls, etc.

Source: S. Brooks et al. 2017. *An Introduction to Privacy Engineering and Risk Management in Federal Systems*. Gaithersburg, Maryland: National Institute of Standards and Technology.

As a result of the likelihood of problems arising from data processing, organizations may be impacted by noncompliance costs, revenue loss from customers canceling products or services, or harm to their external brand reputation or internal culture.

Another challenge that organizations encounter is how to apply a privacy framework effectively in a cloud-computing environment. The top cloud security concerns are data loss and leakage (69%), data privacy (66%), followed by accidental exposure of credentials (44%).[216] To fully implement the privacy frameworks in the cloud requires extending data privacy and protection beyond the organization's network into the software-as-a-service environments—cloud-based applications like Slack, Teams, and Google Workspaces—where employees interact, share information, and where most of the sensitive data is likely found.[217]

A privacy risk assessment—also known as data protection impact assessment or privacy impact assessment (PIA)—is used to identify and evaluate specific privacy risks. It can help organizations weigh the benefits of processing data against the risks; ensure compliance with applicable legal, regulatory, and policy requirements

---

[216] S. Bennett. 2022. Cloud File Security Statistics 2022. *Webinar Care*. 3 October. https://webinarcare.com/best-cloud-file-security-software/cloud-file-security-statistics/.

[217] Azure. 2020. 10 Recommendations for Cloud Privacy and Security with Ponemon Research. *Microsoft Azure* (blog). 29 January. https://azure.microsoft.com/en-us/blog/10-recommendations-for-cloud-privacy-and-security-with-ponemon-research/.; G. Alvarenga. 2022. Cloud Data Security: Securing Data Stored in the Cloud. *CrowdStrike*. 7 June. https://www.crowdstrike.com/cybersecurity-101/cloud-security/cloud-data-security/.; and Principles for Digital Development. How to Secure Private Data Stored and Accessed in the Cloud. https://digitalprinciples.org/resource/howto-secure-private-data-cloud/.

for privacy; and determine an appropriate response. A data protection impact assessment is designed to identify risk arising from the processing of personal data and to allocate resources to mitigating the highest risk and potential damage. The main benefits of privacy risk assessments are that they can serve as an early warning system to detect privacy problems and build safeguards before, not after, major investments; inform decision-making; and signal that the organization is taking privacy concerns seriously through preventive measures.

Depending on potential impacts, organizations may choose to prioritize and respond by

- mitigating the risk (e.g., organizations can apply technical and/or policy measures to the systems, products, or services to minimize the risk to an acceptable degree);
- transferring or sharing the risk (e.g., contracts are a means of sharing or transferring risk to other organizations; privacy notices and consent mechanisms are used to share risk with individuals);
- avoiding the risk (e.g., organizations may determine that the risks outweigh the benefits, and forego or terminate the data processing); or
- accepting the risk (e.g., organizations may determine that problems for individuals are minimal or unlikely to occur, therefore the benefits outweigh the risks, and it is not necessary to invest resources in mitigation).

Data protection regulations, such as California's Consumer Privacy Act (CCPA) or the EU's GDPR do not prescribe an official risk assessment template or specific data protection technologies. Rather, risk identification is required to be part of the process and built into the system's design. Privacy risk can exist throughout the data life cycle, so it is important to manage and govern data end to end.

The purpose of privacy-by-design and privacy-by-default principles is to identify privacy risks and ensure a risk-based approach is applied throughout the entire data life cycle. These concepts, which were developed in the 1990s, can be summarized in seven foundational principles: (i) proactive, not reactive; preventative, not remedial; (ii) privacy as the default setting implies that no action is required by the individual to protect their privacy since it is built into the system; (iii) privacy is integral to the design and functionality of IT systems and business practices; (iv) accommodate all legitimate interests and objectives in a positive-sum "win-win" manner and avoid zero-sum approaches, which entail trade-offs, such as privacy vs. security; (v) embed strong end-to-end security throughout the full data life cycle; (vi) ensure that technology and business practices remain visible and transparent to users and providers and allow for independent verification; and (vii) assign uppermost priority to the interests of the individual user by offering such measures as privacy defaults, appropriate notice, and user-friendly options.[218]

Regulatory authorities, such as the Office of the Australian Information Commissioner, Singapore's Personal Data Protection Commission, or the UK Information Commissioner's Office, have published detailed guides, process maps, and use cases for undertaking privacy impact assessments.[219]

---

[218] A. Cavoukian. 2011. *Privacy by Design The 7 Foundational Principles*.

[219] Office of the Australian Information Commissioner. Guide to Undertaking Privacy Impact Assessments. https://www.oaic.gov.au/privacy/guidance-and-advice/guide-to-undertaking-privacy-impact-assessments.; Personal Data Protection Commission. Guide to Data Protection Impact Assessments. https://www.pdpc.gov.sg/Help-and-Resources/2017/11/Guide-to-Data-Protection-Impact-Assessments.; and UK Information Commissioner's Office. Data Protection Self-Assessment. https://ico.org.uk/for-organisations/sme-web-hub/checklists/data-protection-self-assessment/.

PIAs can help assess (i) the collection of new personal information, (ii) opportunities for data minimization, (iii) anticipated retention periods, (iv) criteria requiring consumer consent, or (v) the utilization of technical and security safeguards.

While the methods for safeguarding these values may vary from case to case, PIA practitioners offer the following suggestions to ensure their effective deployment:[220]

(1) **Do more than a legal compliance check.** Many PIAs are conducted as if they are simply a compliance check against statutory privacy principles without ever asking what impact the activity will have on individuals. For example, the public's visceral reaction against the invasion of their privacy through full body scanners led to their reconfiguration at airports to show screening officers a generic outline of a human body.

(2) **Consider context.** PIAs which focus on one element of a project or program in isolation, rather than the whole ecosystem, will often miss the point. How well users understand the legal protection, transparency, and messaging of a system or project, impacts their level of trust and their ability to take more informed decisions. During the roll out of COVID-19 tracking apps, for instance, many PIAs did not examine the compliance or risks posed by health departments, which were accessing and using the PII collected by the app, yet were operating under a patchwork of privacy laws.

(3) **Test for necessity, legitimacy, and proportionality.** A PIA should not only be about assessing one potential vector for privacy harm, such as compromising personal information, but also whether any negative impacts on individuals are proportionate to the benefits or achievement of the objective. For instance, the Office of the Australian Information Commissioner determined that while 7-Eleven had a legitimate interest in understanding customers' in-store experience, the covert collection of biometrics to achieve that objective was neither necessary nor proportionate to the benefits.

(4) **Test the tech.** If a PIA is not able to test the functionality of a piece of technology as to whether the benefits will *actually* or even *likely* be achieved, no judgment can be made about whether the privacy risks will be outweighed by the benefits.

(5) **Consider customer expectations in gaining trust.** Rather than simply asking "Do you trust this organization/brand?," probing public trust requires a more in-depth set of questions on a case-by-case basis: "Do you trust *this particular way* your data is going to be used *for this particular purpose*, can you see that it will deliver benefits (whether those benefits are personally for you or for others), and are you comfortable that those benefits outweigh the risks for you?"

(6) **Rely on multiple mitigation levers and compare their effectiveness to address privacy risks,** such as technology design and configuration (i.e., choosing which settings to use when implementing off-the-shelf tech); legislation; governance; public communications; user guidance; and staff training.

(7) **Put the recommendations into practice.** Unless findings and recommendations to mitigate privacy risks are taken on board, a PIA will be a pointless exercise in creating a veneer of respectability. Project teams may need time and space to significantly alter the course of a project or abandon it altogether if necessary.

---

[220] A. Johnston. 2022. The Seven Habits of Effective Privacy Impact Assessments. *Salinger Privacy*. 2 August. https://www.salingerprivacy.com.au/2022/08/02/how-to-make-pias-great-again/.

# 6.3. The Data Processing Ecosystem

A key factor in the management of privacy risk is the organization's role(s) in the larger data processing ecosystem, which can affect not only its own legal obligations, but also the measures it may take in relation to how other entities in the ecosystem are managing privacy risk. As illustrated in Figure 22, the ecosystem consists of all individuals, communities, and organizations, as well as their relationships with each other, that are stewarding data, creating value from it, or are affected by any of those activities.

The NIST Privacy Framework provides a common language to communicate privacy requirements with entities within the data processing ecosystem. For example, an organization that makes decisions about how to collect and use data about individuals may use a custom profile to express privacy requirements to an external service provider (e.g., a cloud provider to which it is exporting data); and the external service provider as the data processor may then use this profile to demonstrate the measures it has adopted to process data in line with contractual obligations. Similarly, a manufacturer may use a target profile to determine the capabilities to build into its products so that its customers can meet the privacy needs of the end user.

**Figure 22:** The Ecosystem for Data Processing

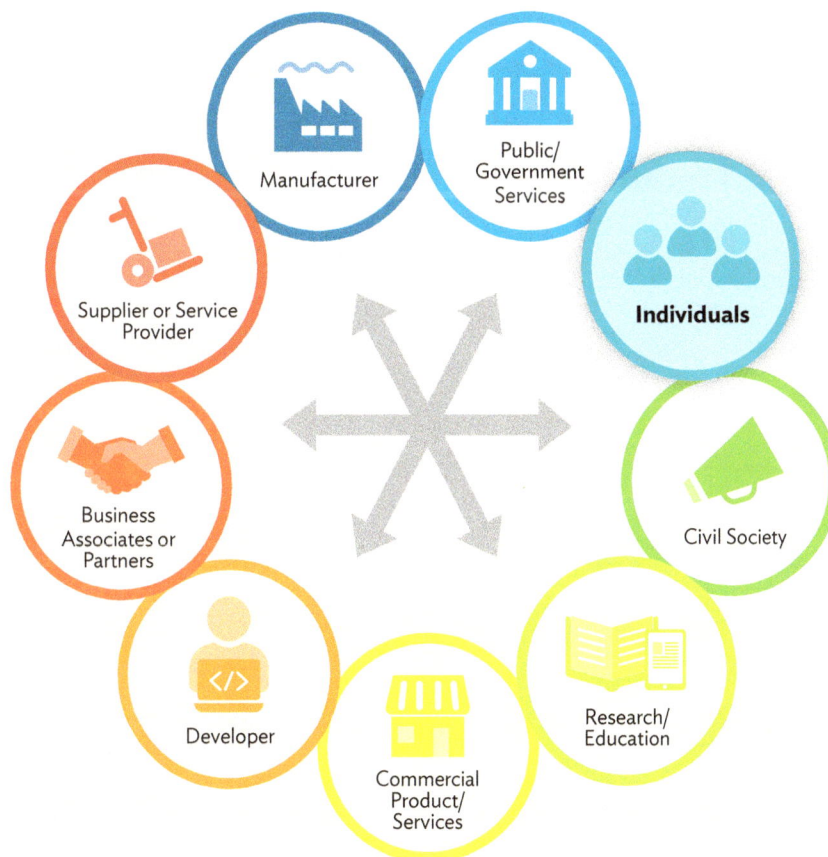

Source: NIST Privacy Framework.

To function effectively, this ecosystem of organizations needs to be based on a strong data infrastructure that relies on sustainable funding, is trusted and trustworthy, and is geared to draw maximum value from its data assets. Data infrastructure can be hard to visualize and understand. A data ecosystem map can help to develop a shared understanding of how participants create value from data; identify gaps and opportunities to improve its functioning (e.g., by changing how data is accessed, used, and shared); identify potential users and communities who might benefit from the creation of a new data infrastructure in a sector; and help set up institutions, standards, or data access initiatives that could add value and support data flows Figure 23 visualizes two data ecosystem maps for a manufacturing supply chain and a fintech sandbox.[221]

## Figure 23: Mapping Data Ecosystems

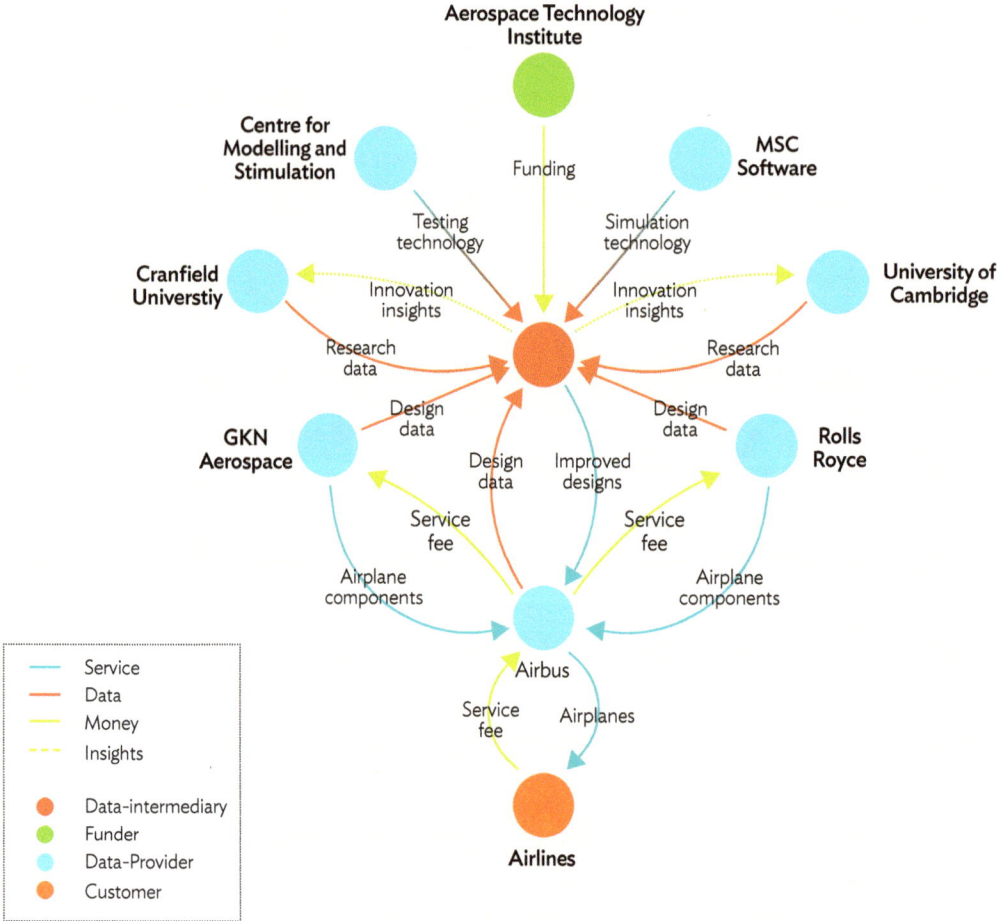

Airbus is using the APROCONE product design platform to share engineering data securely, significantly reducing time and cost spent in the design process and optimizing its supply chain

*continued on next page*

---

221 J. D'Addario. 2022. Mapping Data Ecosystems: Methodology. *Open Data Institute*. 1 April. https://theodi.org/article/mapping-data-ecosystems/.

**Figure 23** *continued*

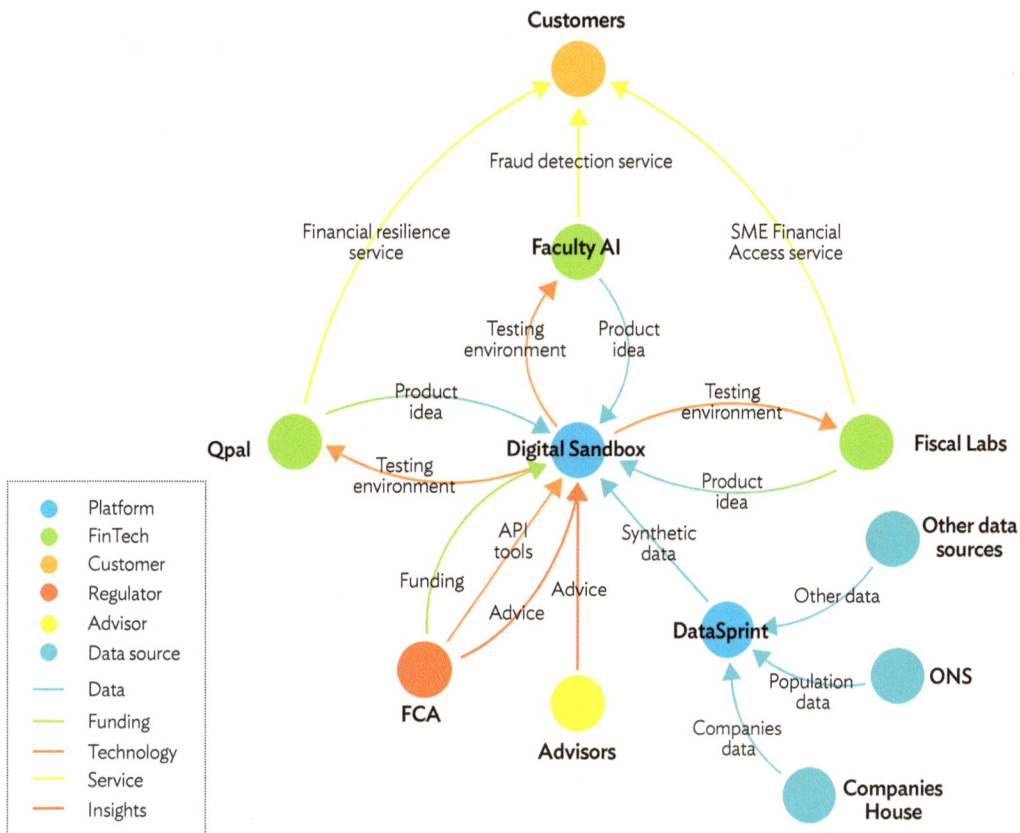

This data ecosystem map shows how the Digital Sandbox innovation platform uses synthetic data, expert advice, and code testing technology to help fintechs improve their products and services.

Source: Open Data Institute, Mapping Data Ecosystems: Methodology, 2022.

# 6.4. Establishing Data Governance

Data governance involves a comprehensive set of rules, policies, and procedures that guide how organizations manage, use, and share data internally and across data ecosystems.

The OECD defines data governance as principles and policy guidance on how governments can maximize the cross-sector benefits of all types of data—personal, non-personal, open, proprietary, public, and private—while protecting the rights of individuals and organizations.[222] The World Bank notes that data governance consists of four tasks: strategic planning, developing rules and standards, developing mechanisms of compliance and enforcement, and generating the learning and evidence needed to gain insights and address policy challenges (Box 5).[223]

---

[222]  OECD. Data Governance. https://www.oecd.org/sti/ieconomy/enhanced-data-access.htm.
[223]  World Bank. World Development Report 2021: Data for Better Lives.

## Box 5: The Pillars of an Integrated National Data Ecosystem

An integrated data ecosystem should provide innovative answers to fundamental questions: How does a country extract value from data? How is trust in data ensured? What data governance structures are necessary to oversee and share data? What resilient data infrastructure is needed to manage data securely? What skills do citizens and businesses need to benefit from data?

The World Bank's 2021 World Development Report, *Data for Better Lives*, called for a new social contract for data, founded on value, trust, and equity. It argued that data governance is key to realizing the value of data through the trustworthy and equitable production, use, and reuse of data to support those areas of economic activity crucial for a country's sovereignty, prosperity, and global competitiveness.

Delivering on this promise means building an integrated national data ecosystem that
- provides unified, transparent, and secure access to trustworthy, high-quality data, including administrative, statistical, business, industrial, scientific, and real-time, machine-generated data;
- consists of a set of interconnected, sector-specific trusted data platforms which are interoperable and where (a) personal and non-personal data are secure; (b) stakeholders can freely exchange data, subject to rules that guarantee data privacy; (c) users have fast, reliable access to relevant information and services based on international and national data standards; and (d) interoperability frameworks are in place;
- is implemented with reliable digital infrastructure, including data stacks, ethical, analytical tools (artificial intelligence and/or machine learning), enhanced cybersecurity, and pooled, distributed infrastructure (e.g., cloud, edge, the Internet of Things, and networks) for managing and processing data to create value;
- is underpinned by laws and regulations to establish trust and safeguard users' rights, facilitated by economic policies to support innovation, and overseen by inclusive governance institutions, which ensure the quality, accessibility, protection, availability, reusability, and preservation of the country's unstructured and structured data and associated metadata; and
- is supported by human capital policies to enhance evidence-based policymaking and the use of data in the public sector and to increase digital skills, cybersecurity awareness, and data literacy for all.

Source: D. Deasy, Y. Eferin, and O. Petrov. 2022. Integrated National Data Ecosystems: The next Stage of Digital Transformation. *World Bank* (blog). 6 September.

---

Although data governance is an important component of 21st century governance, researchers and policymakers are grappling with what a comprehensive approach to data governance might look like. The Global Data Governance Mapping Project[224] at George Washington University has proposed six attributes to govern data in a comprehensive manner (Figure 24):

(1) vision and strategies for data in the economy and society;
(2) legal and regulatory regimes for data types and uses;
(3) responsible ethical and human rights guidelines for data use and re-use;
(4) changes to institutional structures in response to data-driven transformation;
(5) public participation; and
(6) mechanisms for international cooperation.

---

[224] T. Struett, A. Zable, and S. Ariel. 2022. *Global Data Governance Mapping Project: Year Two Report*.

## Figure 24: Global Data Governance Mapping

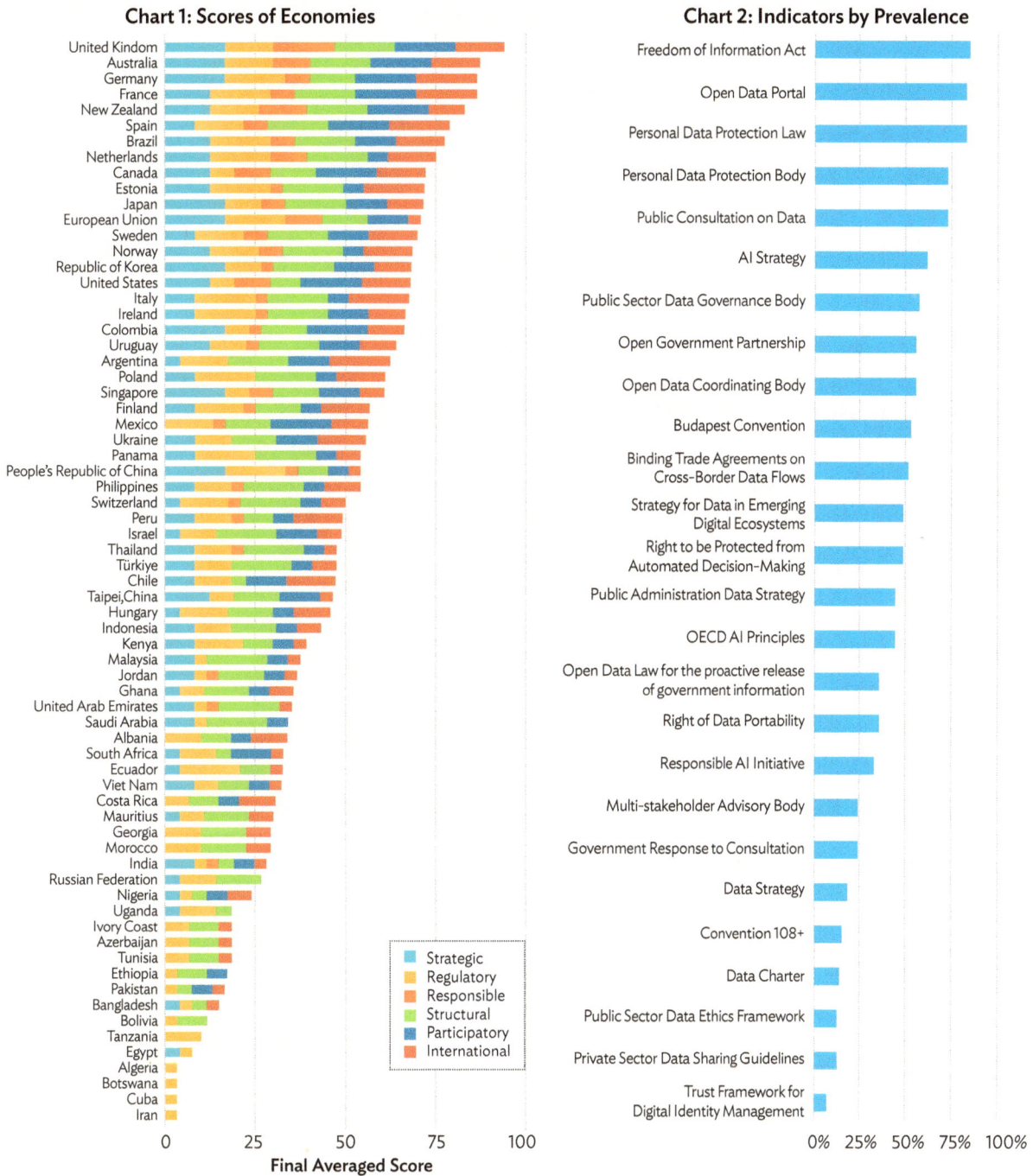

### Chart 1: Scores of Economies

Economies (top to bottom): United Kindom, Australia, Germany, France, New Zealand, Spain, Brazil, Netherlands, Canada, Estonia, Japan, European Union, Sweden, Norway, Republic of Korea, United States, Italy, Ireland, Colombia, Uruguay, Argentina, Poland, Singapore, Finland, Mexico, Ukraine, Panama, People's Republic of China, Philippines, Switzerland, Peru, Israel, Thailand, Türkiye, Chile, Taipei,China, Hungary, Indonesia, Kenya, Malaysia, Jordan, Ghana, United Arab Emirates, Saudi Arabia, Albania, South Africa, Ecuador, Viet Nam, Costa Rica, Mauritius, Georgia, Morocco, India, Russian Federation, Nigeria, Uganda, Ivory Coast, Azerbaijan, Tunisia, Ethiopia, Pakistan, Bangladesh, Bolivia, Tanzania, Egypt, Algeria, Botswana, Cuba, Iran

Legend: Strategic, Regulatory, Responsible, Structural, Participatory, International

X-axis: Final Averaged Score (0, 25, 50, 75, 100)

### Chart 2: Indicators by Prevalence

Indicators (top to bottom): Freedom of Information Act, Open Data Portal, Personal Data Protection Law, Personal Data Protection Body, Public Consultation on Data, AI Strategy, Public Sector Data Governance Body, Open Government Partnership, Open Data Coordinating Body, Budapest Convention, Binding Trade Agreements on Cross-Border Data Flows, Strategy for Data in Emerging Digital Ecosystems, Right to be Protected from Automated Decision-Making, Public Administration Data Strategy, OECD AI Principles, Open Data Law for the proactive release of government information, Right of Data Portability, Responsible AI Initiative, Multi-stakeholder Advisory Body, Government Response to Consultation, Data Strategy, Convention 108+, Data Charter, Public Sector Data Ethics Framework, Private Sector Data Sharing Guidelines, Trust Framework for Digital Identity Management

X-axis: 0%, 25%, 50%, 75%, 100%

Note:
**Chart 1** shows the United Kingdom, Australia, New Zealand, and France take the most comprehensive approach to data governance at national and international levels. High-income economies focus their data governance efforts more the international (red) and responsible (orange) atributes, while lower middle-income economies tend to concentrate on structural (dark blue) and regulatory (yellow) actions. Although most economies seek public comment on proposed data laws and regulations, there is limited evidence that policymakers revise their data governance policies in response to public concerns.

**Chart 2** illustrates which indicators are the most prevalent among the ~70 sample economies. Over 75% have Freedom of Information Acts, open data portals, and personal data protection laws. In contrast, fewer than 25% have a data strategy, a public sector data ethics framework, private sector data sharing guidelines, or trust framework for digital identity management.

Source: Global Data Governance Mapping Project. 2022

In many low- or middle- income countries implementation of new data privacy laws is challenging due to the following:

- The complexity and uncertainty about how to act in accordance with data protection laws, or how to act in their absence, present major barriers to data sharing. The absence of national data privacy laws can derail data-intensive projects as donor organizations or investors may be wary of providing support in jurisdictions that lack such protection.
- There is a lack of legal skills and knowledge needed to comply with data privacy laws across government.[225] Government agencies that want to collaborate with third parties must have the ability to negotiate complex contracts that prevent their data from being reused for unintended purposes, and they need the capability to monitor these contracts to ensure accountability.
- There is a shortage of funding for data protection agencies (DPAs). Even in countries with well-established data protection and privacy regimes, DPAs responsible for overseeing these efforts often lack sufficient budget, staffing or authority.

Several lower middle-income countries are piloting new approaches for data protection and privacy law by modifying their approaches to meet domestic priorities and, in some cases, extending protections beyond those offered by the GDPR. Kenya's Data Protection Act 2019, for instance, requires organizations to conduct data protection risk assessments for activities that have the potential of creating high risk to the rights and freedoms of citizens (e.g., credit scoring, CCTV monitoring in public spaces, DNA processing),[226] which go beyond GDPR requirements.

Some countries, especially in Asia, have adopted policy and regulatory features that deepen and extend the reach of the state. A recent review by the Carnegie Endowment for International Peace compares emerging data governance models between India and the Republic of Korea.[227] The study counters the argument that the world faces a simple binary choice between democratic models or authoritarian ones and demonstrates that countries are pioneering their own approaches, combining elements of their unique democratic institutional frameworks with national requirements and policies derived from distinctive political cultures.

Progress on data governance has been uneven as different government agencies have conflicting policy goals, making it difficult to develop a clear, consistent vision and strategy. This can result in missed opportunities to share, combine, and use data to solve problems in the public and private sectors. India's Digital Personal Data Protection Bill 2023, for instance, introduces a broader definition of personal and sensitive personal data and does not provide for the right to data portability and the right to be forgotten. Its 2022 draft came under criticism for giving too many powers to the state, imposing a high compliance

---

225 Paris 21. 2020. Paris21 Launches Capacity Development 4.0 Guidelines. https://paris21.org/news-center/news/paris21-launches-capacity-development-40-guidelines.

226 G. Mutung'u and F. Ogonjo. 2021. *Simplified Data Protection Impact Assessment for Small Organisations*. Nairobi: Strathmore University Center of Intellectual Property and Technology Law. https://cipit.strathmore.edu/wp-content/uploads/2021/05/Simplified-DPIA.pdf.

227 E. Feigenbaum and M. Nelson. 2022. Data Governance, Asian Alternatives: How India and Korea Are Creating New Models and Policies. *Carnegie Endowment for International Peace*. 31 August. https://carnegieendowment.org/2022/08/31/data-governance-asian-alternatives-how-india-and-korea-are-creating-new-models-and-policies-pub-87765.

burden, and introducing extensive data localization requirements.[228] In response, the government has made amendments and both houses of Parliament have since passed the revised bill,[229] which will take effect after the presidential sign-off.

Strong country ownership is required for effective data governance reforms and institutions charged with their implementation need to be granted the necessary authority and funding to implement them. For policymakers to see value in data governance, they must first see data as an asset whose maintenance is worth investing in—a view they are more likely to hold if they are keen on data-driven decision-making.

Beyond raising awareness of the risks of data misuse, policymakers need to understand the importance of responsible data use. Three benefits are relevant: first, clear rules and reliable data protection mechanisms can help policymakers achieve their objectives by facilitating data access, sharing, and re-use in a safe and transparent manner; second, those rules can be implemented at reasonable costs, both in terms of government resources and compliance burden; and third, universal principles of data protection and privacy can be applied in a manner that is tailored to local realities and is proportional to risk.

Because the digital divide closely mirrors existing inequalities, there is a significant risk that already disadvantaged groups will fall further behind unless steps are taken to ensure that the benefits of digitalization are fairly distributed across the whole of society. Despite efforts by community groups over the last 10 years, the value of data—defined in terms of actionable insights that can be used to improve services and policy—does not yet flow back to the individuals, communities, and organizations who initially provided the underlying data ("open loop").

## 6.5. The Need for Regional and Global Solutions

Nascent global debates about data governance have focused largely on issues of cross-border data transfers, and often reflect the priorities of wealthier countries and multinational corporations that are keen to defend their business interests. As a result, the data policy landscape remains fragmented, with the world's most powerful blocks taking different approaches.

The lack of global consensus on how countries should govern data, especially regarding data privacy and protection, makes it harder for individual countries to determine the right path. In principle, a global agreement on a single set of data privacy principles already exists today in the form of the OECD *Guidelines on the Protection of Privacy and Transborder Flows of Personal Data* and *Convention*. But agreement on actual implementation guidance that would provide the necessary protection remains elusive, even if such guidance could be kept flexible so countries could apply principles in a manner that is consistent with their needs, priorities, and capacities.

---

[228] A. Burman. 2022. The Withdrawal of the Proposed Data Protection Law Is a Pragmatic Move. Carnegie India. 22 August. https://carnegieindia.org/2022/08/22/withdrawal-of-proposed-data-protection-law-is-pragmatic-move-pub-87710.

[229] PRS Legislative Research. The Digital Personal Data Protection Bill, 2023. https://prsindia.org/billtrack/digital-personal-data-protection-bill-2023.

Global institutional arrangements to bring together data protection officials as a unified body also remain underdeveloped. There is a need for meaningful multilateral engagement between DPAs and organizations focused on supporting economic growth and better social outcomes. The most significant development to date is the creation of the Global Privacy Assembly 2019, which is crafting guidance on privacy policy and advocating on wider digital policy issues.[230]

Regionally, various forums and networks exist which foster knowledge sharing and collaboration. The Asia Pacific Privacy Authorities, a forum of privacy and data protection authorities which stretches into Latin America, was formed in 1992 and meets biannually.[231] The Association of Southeast Asian Nations has created a Data Protection and Privacy Forum, a framework on digital data governance, a data management framework, and model standard clauses for cross-border data flows.[232] Privacy and data protection in the Asia-Pacific Economic Cooperation member states can exchange information via the Cross-border Privacy Enforcement Arrangement[233] which promotes cooperation. AsiaDPO is a peer-to-peer community of privacy and data protection professionals from different disciplines and backgrounds, registered as a society in Singapore.[234]

In South Asia, most digital policy research has focused on India, including work by the Data Governance Network at the IDFC Institute.[235] A regional perspective on hard and soft infrastructure is being offered by LIRNEasia, a Colombo-based policy and regulation think tank.[236] The Future of Privacy Forum Asia-Pacific has been active in the region since 2021, publishing research and providing a platform for knowledge exchange and networking among practitioners, with the objective to work toward convergence of data protection regulations and best privacy practices in the Asia and Pacific region.[237]

## 6.6. Plausible Data Futures

Data-rich organizations such as banks, health-care firms, utilities, and major manufacturers and retailers are under growing pressure to change their data governance and management practices.

By some measure, up to 90% of current IT budgets are spent trying to manage internal complexities, crowding out space for data innovation to improve productivity or customer experience. Many large organizations experience internal tensions over the management of customer data. A frequent tension exists between the roles of the chief information officer, whose responsibility is to collect, encrypt, and secure data; the chief digital

230  Global Privacy Assembly. List of Accredited Members. https://globalprivacyassembly.org/participation-in-the-assembly/list-of-accredited-members/.
231  Asia Pacific Privacy Authorities Forum. About. https://www.appaforum.org/about/.
232  Personal Data Protection Commission Singapore. ASEAN Data Management Framework and Model Contractual Clauses on Cross Border Data Flows. https://www.pdpc.gov.sg/Help-and-Resources/2021/01/ASEAN-Data-Management-Framework-and-Model-Contractual-Clauses-on-Cross-Border-Data-Flows.
233  Asia Pacific Economic Cooperation. APEC Cross-border Privacy Enforcement Arrangement. https://www.apec.org/About-Us/About-APEC/Fact-Sheets/APEC-Cross-border-Privacy-Enforcement-Arrangement.
234  AsiaDPO. About. https://www.asiadpo.org/.
235  IT for Change. Data Governance Network. https://itforchange.net/data-governance-network-community-data-ownership.
236  LIRNEasia. https://lirneasia.net/.
237  Future of Privacy Forum. FPF Asia Pacific. https://fpf.org/fpf-asia-pacific/.

officer, who is tasked with pushing data out to entice and advocate for new users; and the chief data officer, who plays a risk, compliance, and business role.[238] For organizations with expansive data collection operations, additional complexities arise from multiple legacy systems, a web of multilayered data-sharing agreements, and, quite often, an ongoing lack of clarity on how to integrate data into their businesses.

One plausible future was suggested by professors Pentland and Rahnama, who argued that companies will have to reorganize their data operations around the new fundamental rules of consent, insight, and flow:[239]

**Rule 1: Trust over transaction.** The ever-expanding digital landscape implies that organizations must broaden their trust capabilities around security, privacy, and cyber. As consumers are becoming more aware of their identity rights, basing their decisions of giving consent on values, and demanding the ethical use of data and responsible AI, organizations need to consistently cultivate trust and explain in common-sense terms how their data is being used and what is in it for them. Organizations face the additional complexity of having to negotiate and secure trust across an entire ecosystem of technologies (e.g., wired home, virtual assistants, communications, mobility devices).

As organizations will be forced to put identity and trust management at the core of their customer experience and business process, this will be a departure from the current practice of accumulating as much data as possible, transaction by transaction. With the shift toward greater customer control, data collected with *meaningful* consent will soon be the most valuable data source, because it is the only data that organizations will be allowed to work with. Data cooperatives could provide users with new options for data sharing and secure each user's consent for the option they are most comfortable with.[240]

**Rule 2: Insight over identity.** Currently, companies routinely transfer large amounts of personally identifiable information (PII) through a complex web of data agreements, compromising both privacy and security. Today's data architecture, such as federated learning, trust networks,[241] and identity management,[242] already makes it possible to gain insight from data without having to acquire or transfer the data itself. The premium will be on IT's ability to enable innovation, requiring a shift away from its traditional role as protector of digital assets to that of a data diagnostician and purveyor of small blocks of code using low-code or no-code platforms to create required data products. This IT-as-a-service approach puts the product at the center of the operating model. The co-design of algorithms and data can facilitate this process of insight extraction by structuring each to better meet the needs of the other. As a result, rather than moving data around, the algorithms exchange non-identifying statistics instead. Australia's DSpark, for instance, aggregates and anonymizes over

---

[238] For a perspective on the tenuous position of chief data officers, see T. Davenport, R. Bean, and J. King. 2021. Why Do Chief Data Officers Have Such Short Tenures? *Harvard Business Review*. 18 August. https://hbr.org/2021/08/why-do-chief-data-officers-have-such-short-tenures.

[239] H. Rahnama and A. Pentland. 2022. The New Rules of Data Privacy. *Harvard Business Review*. 25 February. https://hbr.org/2022/02/the-new-rules-of-data-privacy.

[240] S. Mehta, M. Dawande, and L. Mu. 2022. This Is the Key to Designing Sustainable Data Cooperatives. *World Economic Forum*. 1 February. https://www.weforum.org/agenda/2022/02/the-key-to-designing-sustainable-data-cooperatives/.

[241] A. Brundyn, E. Scoullos, and D. Williams. 2022. Using Federated Learning to Bridge Data Silos in Financial Services. *NVIDIA Developer* (blog). 16 August. https://developer.nvidia.com/blog/using-federated-learning-to-bridge-data-silos-in-financial-services/.; and N. Cheong. 2022. Design a Federated Learning System in Seven Steps. *Integrate.ai*. 16 June. https://www.integrate.ai/blog/design-a-federated-learning-system-in-seven-steps-pftl.

[242] Tech Target. What Is Federated Identity Management (FIM)? How Does It Work? https://www.techtarget.com/searchsecurity/definition/federated-identity-management.

1 billion mobility data points every day, which it then turns into marketable intelligence on everything from demographics to shopping habits for other companies—without ever selling or transferring the data itself.[243]

**Rule 3: Flows over silos.** Once companies can obtain customer data based on meaningful consent (Rule 1) and acquire insights without transferring data (Rule 2), this should open new possibilities for internal data teams (chief information officers, chief digital officers, chief information security officers, etc.) to collaborate to facilitate the flow of insights. Take the case of a bank's mortgage unit which secures a customer's consent to expedite their relocation by sharing the new address with service providers such as moving companies, utilities, and internet providers. With the bank acting as a broker or agent to provide personalized services for customers, the result is a data ecosystem that is trustworthy, secure, and under customers' direct control.

The emergence of data representatives, agents, and custodians makes it possible to manage consent at scale, serving as trusted hubs for users' personal data and acting as their agents in the marketplace. The key value proposition of the new data economy built on trust, insight, and flow is that wealth generation can be distributed better and more equitably while reducing privacy and security risks. Rather than hoarding their data assets, people can take advantage of new opportunities such as getting a financial return on their data or better personalized services.

# 6.7. Areas for Future Research and Collaboration

The data-for-development domain presents promising areas for new research and engagement (Box 6):[244]

- What new models and frameworks can be used to examine data protection and privacy through the lens of economic development? This line of inquiry should be of particular interest to policymakers in countries that are modifying their approach to data protection and privacy to meet domestic needs and priorities, including supporting economic growth and innovation.
- What are key drivers for successful data partnerships? Which capacity-building programs are helpful in preparing organizations for these new arrangements?
- How can greater transparency and evaluation be brought to public–private data partnerships? What type of monitoring and evaluation expertise is needed to implement data-intensive projects, including risk assessments and ways to mitigate risk?
- What are the options for greater cooperation between data privacy and international development communities?

243   DSpark. https://www.dspark.com.au/.
244   Center for Global Development. Governing Data for Development: Trends, Challenges, and Opportunities. https://www.cgdev. org/publication/governing-data-development-trends-challenges-and-opportunities.

**Box 6:** The Development Data Partnership: Using Third-Party Data for Public Good

The Development Data Partnership is a collaboration between international organizations and technology companies to facilitate the efficient and responsible use of third-party data in international development. The partnership brings international organizations—including the World Bank, Asian Development Bank, European Bank for Reconstruction nd Development, Inter-American Development Bank, International Monetary Fund, Organisation for Economic Co-operation and Development, United Nations Development Programme , and the Rockefeller Foundation— together to collaborate with private sector companies for the public good. The private sector partners provide data and/or services and can open markets in emerging economies, discover new data methods that inform future products, and increase staff skills through collaboration and secondments.

Some recent examples of their work include the following:

- Empowering female entrepreneurs to cross over into male-dominated sectors
- Green jobs for women and diverse groups to promote equality and climate resilience
- Using Big Data and machine learning to strengthen road safety
- Analyzing the evolution of telework in Latin America and the Caribbean
- Mapping the effects of city growth on urban mobility and traffic patterns
- Identifying and mitigating the socioeconomic effects of heat waves

AtlasAI  EARTHMETRY  esa  esri  GitHub  Google  homebase  K khalti  Linked in

mapbox  Mapillary  Meta  Microsoft  OOKLA  Outlogic  Orbital Insight  plume labs  PREMISE

QUADRANT  SPACEKNOW  spectus  STARLINK  tomorrow  veraset  WhereIsMyTransport  waze

Source: Development Data Partnership.

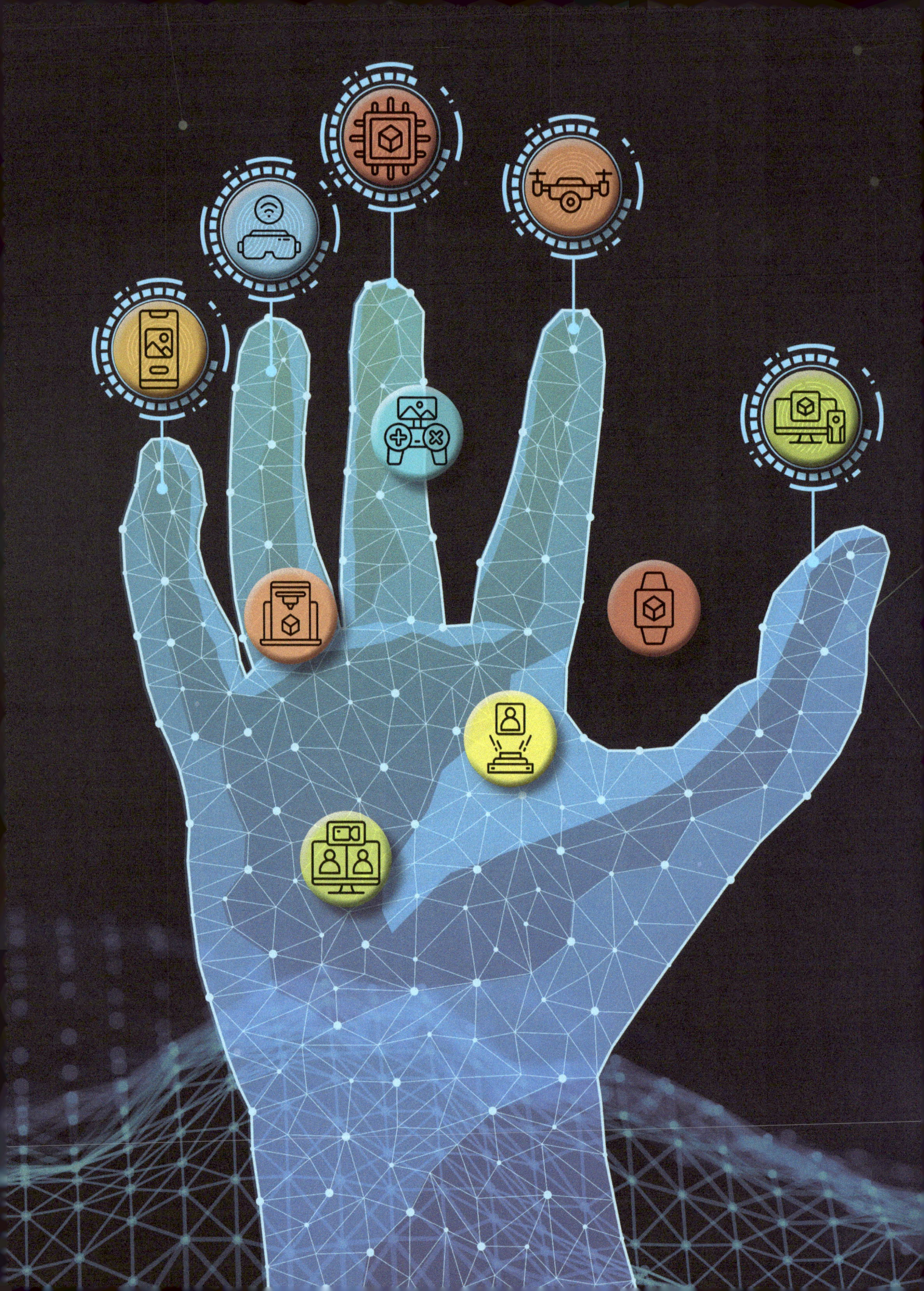

# 7

ETHICAL AI RISKS

Digital transformations promise efficiency gains and new services, products, and business opportunities, freeing up resources for activities that require human touch, creativity, empathy, or decisions that are hard or impossible to model. AI and large language models are playing a significant role in these digital transformations. Generative AI applications can create new content in the form of text, code, voice, images, videos, and processes in response to user requests and are becoming widely available through products such as ChatGPT from OpenAI and BARD from Google. ChatGPT surpassed over 100 million monthly users within 2 months of its public launch in November 2022, setting a world record as the fastest-growing web application.

Amid the hype, leaders in tech, industry, academia, and civil society are issuing ever-more urgent warnings about the risks of AI amid calls for a regulatory response by governments.[245]

What is widely seen as an inflection point for digital transformation requires a holistic ethical framing to avoid unwanted developments such as undermining privacy, creating unfair outcomes, or losing human control over increasingly intelligent systems. Another pressing ethical dilemma concerns malicious and/or dual-use applications of generative AI and large language models.[246] From autonomous weapons systems to the conversion of civilian disaster response robots for military use to unauthorized use of facial recognition technology and unsupervised decision-making algorithms, the dual-use nature of AI technology can entail significant security risks to individuals, governments, industries, and the future of humanity.[247] Reflecting widespread concerns about the lack of guardrails for responsible AI research, Google DeepMind recently submitted a proposal for an early warning system against novel AI risks with dangerous capabilities.[248]

The following sections will compare recent developments in AI regulations in the PRC, Europe, and the US; outline a field-tested approach for assessing and mitigating ethical AI risks; and provide an overview of AI tool kits and guidelines intended to tackle the broad set of rapidly evolving legal, ethical, and societal challenges associated with AI.

# 7.1   The Global Race to Regulate Artificial Intelligence

The PRC has taken an early lead in adopting consumer-facing AI regulations.[249] In March 2022, the PRC's Cyberspace Administration passed a regulation governing companies' use of algorithms in online recommendation systems, requiring that such services are moral, ethical, accountable, transparent, and

[245] Center for AI Safety. Statement on AI Risk. https://www.safe.ai/statement-on-ai-risk.; J. Nelson. 2023. Take AI Warnings Seriously, Says UN Secretary-General. *Decrypt*. 14 June. https://decrypt.co/144692/take-ai-warnings-seriously-un-secretary-general/. ; and M. Sullivan. 2023. What Is the Real Point of All These Letters Warning about AI? *Fast Company*. 31 May. https://www.fastcompany.com/90902786/what-is-the-real-point-of-all-these-letters-warning-about-ai.

[246] Malicious AI Report. The Malicious Use of Artificial Intelligence. https://maliciousaireport.com/.

[247] Some researchers have suggested the implementation of a Dual Use Research of Concern framework, originally developed for biological research, for generative AI and Large Language Models. A. Grinbaum and L. Adomaitis. Dual Use Concerns of Generative AI and Large Language Models. *arXiv*. 13 May. https://doi.org/10.48550/arXiv.2305.07882.

[248] T. Shevlane. 2023. An Early Warning System for Novel AI Risks. *Google Deep Mind* (blog). 25 May. https://www.deepmind.com/blog/an-early-warning-system-for-novel-ai-risks.

[249] C. Howell. 2022. AI Regulation: Where Do China, the EU, and the U.S. Stand Today? *The National Law Review*. 3 August. https://www.natlawreview.com/article/ai-regulation-where-do-china-eu-and-us-stand-today.; Global Legal Insights. AI, Machine Learning & Big Data Laws and Regulations: China. https://www.globallegalinsights.com/practice-areas/ai-machine-learning-and-big-data-laws-and-regulations/china.

"disseminate positive energy." The regulation mandates companies to notify users when an AI algorithm is playing a role in determining which information to display to them and give them the option to opt out of being targeted. Additionally, the regulation prohibits algorithms that use personal data to offer different prices to consumers.[250] Most recently, the PRC's internet regulator released a draft proposal for regulating generative AI systems.[251] The draft strongly favors protecting society (and the government) from risks posed by AI systems and tech companies, rather than applying a laissez-faire governance approach that would give the private sector substantial latitude in developing new AI products. On the question of whether developers or the government should take the lead in defining rules for AI, the regulation squarely shifts power to the regulators.[252]

In September 2022, the Shenzhen city government passed the PRC's first local level AI regulation in pursuit of becoming the country's AI and tech hub, encouraging government organizations to be the forerunners in utilizing AI technology. Shenzhen-based AI services and products that are assessed as "low risk" can be piloted even in the absence of local and national norms if they adhere to international standards. In the same month, Shanghai passed the PRC's first provincial-level law covering AI development through a grading system and supervised "sandboxes." One highlight of the Shanghai AI regulation is that it stipulates a certain degree of tolerance for minor infractions to encourage exploration of scientific frontiers and inspiring innovation.

The European Commission has made it a priority to create a far-reaching regulatory framework that would prevent and minimize AI's negative effects and protect the "right to non-discrimination, freedom of expression, human dignity, personal data protection, and privacy." The proposed approach targets the riskiest AI systems which, if passed into law, could set a new global standard for AI oversight.

The ambitious draft AI Act of 2021,[253] which was approved by the EU Parliament on 14 June 2023 and is awaiting ratification by national governments before likely going into effect by end 2023,[254] seeks to regulate the entire industry ("horizontal legislation") and takes a risk-based approach, with different requirements and obligations based on the level of risk:

- At the top of the risk pyramid are three AI systems that would be banned outright as EU lawmakers consider them "unacceptable" risks to public safety and human rights by (i) deploying cognitive behavioral manipulation of people or specific vulnerable groups, such as children or the older people, resulting in physical or psychological harm; (ii) opening the possibility of

250  Y. Wu. 2022. AI in China: Regulations, Market Opportunities, Challenges for Investors. *China Briefing.* 14 October. https://www.china-briefing.com/news/ai-in-china-regulatory-updates-investment-opportunities-and-challenges/.
251  R. Iyengar. 2023. The Global Race to Regulate AI. *Foreign Policy* (blog). 5 May. https://foreignpolicy.com/2023/05/05/eu-ai-act-us-china-regulation-artificial-intelligence-chatgpt/.
252  H. Toner et al. 2023. How Will China's Generative AI Regulations Shape the Future? A DigiChina Forum. *Stanford University DigiChina Project.* 19 April. https://digichina.stanford.edu/work/how-will-chinas-generative-ai-regulations-shape-the-future-a-digichina-forum/.
253  EUR-Lex. Proposal for a Regulation of the European Parliament and of the Council Laying Down Harmonised Rules on Artificial Intelligence (Artificial Intelligence Act) and Amending Certain Union Legislative Acts. https://eur-lex.europa.eu/legal-content/EN/TXT/?uri=celex%3A52021PC0206.
254  A. Satariano. 2023. Europeans Take a Major Step Toward Regulating A.I. *The New York Times.* 14 June. https://www.nytimes.com/2023/06/14/technology/europe-ai-regulation.html.

"social scoring" by classifying people based on behavior, socioeconomic status, or personal characteristics; and (iii) applying them for real-time and remote biometric identification, such as facial recognition, by law enforcement in public spaces.

■ The next level involves "high-risk" AI systems that could create adverse impacts on fundamental rights and safety but may be authorized subject to strict regulation, including the EU's product safety legislation. These systems and applications, which may account for 5% to 15% of the AI universe fall into eight specific areas that will have to be registered in an EU database: biometric identification and categorization of natural persons; management and operation of critical infrastructure; education and vocational training; employment, work management, and access to self-employment; access to essential private services and public services; law enforcement; migration, asylum, and border control management; and assistance in legal interpretation and application of the law. Additionally, all safety-critical AI systems (e.g., in cars, planes, medical devices, toys) would be rated as high risk. Providers would have to meet strict conformity requirements, such as high-quality training; validation of testing data; traceability and auditability; transparency to end users; human oversight (e.g., a "kill switch" if the AI system needs to be shut down); and robustness, accuracy, and cybersecurity of the underlying AI system.

■ Generative AI would have to comply with transparency requirements, disclosing that the content was generated by AI, a design that is intended to prevent it from generating illegal content, and publishing summaries of copyrighted data used for training the AI.

■ "Limited risk" or "minimal risk" systems are allowed to operate with less restrictions, but would be subject to transparency requirements, such as notifying people when they are interacting with an AI system or applying labels to deepfakes. Examples include chatbots, spam filters, or inventory management systems. Big-tech platforms would be able to avoid significant disturbance to their business as AI-based systems in social media, search, online retailing, or app stores are not deemed high risk.

The most significant impact of these new rules is that they will apply not only to providers of AI systems based in the EU, but also to businesses outside the EU that provide AI-based products and services used inside the EU—effectively setting a global standard that will force tech companies and those embedding AI systems in their products and services to adapt.[255] From a governance and enforcement perspective, the AI Act would establish a European AI Board to oversee rulemaking. Noncompliance could lead to substantial fines of up to 6% of annual worldwide sales.

As the bill is winding its way through national parliaments and public consultations, several design challenges have surfaced.[256] The AI Act requires, for instance, that data sets must be error free and that humans need to be able to fully understand how AI systems work—a challenge given the complexity of today's neural networks. Tech companies are also wary of giving external auditors or regulators access to their models, source code, and algorithms to comply with the law. There is also controversy over whether the AI Act should ban the use of facial recognition outright if it involves national security

255  A. Engler. 2022. The EU AI Act Will Have Global Impact, but a Limited Brussels Effect. *Brookings*. 8 June. https://www.brookings.edu/articles/the-eu-ai-act-will-have-global-impact-but-a-limited-brussels-effect/.
256  M. Heikkilä. 2022. A Quick Guide to the Most Important AI Law You've Never Heard of. *MIT Technology Review*. 13 May. https://www.technologyreview.com/2022/05/13/1052223/guide-ai-act-europe/.

or law enforcement. Another debate revolves around the classification of "high-risk" AI systems (e.g., systems for lie detection or allocation of social assistance payments), with one side arguing that the bill does not offer enough protection for people, while the other side warns that the vast scope of regulation will become a brake for innovation.

Other critics[257] have noted that the proposal takes a simplistic view of how AI systems function by arguing the following: (i) AI is neither a product nor a "one-off" service that can be subjected to existing EU consumer safety rules, but more akin to a system that is dynamically delivered though a complex web of procurement, outsourcing, reuse of data, etc. From a liability perspective, difficulties may arise from determining the distribution of sole and shared responsibility. The complexity, autonomy, and opacity of certain AI systems may make it difficult to explain their inner functioning in practice ("black box"). (ii) Those impacted by AI systems (i.e., end users) have no rights and almost no role in the proposed AI Act. (iii) The categorization of AI risks is arbitrary and not based on objective criteria. (iv) The proposed act lacks an impact risk assessment of fundamental rights that would apply for all AI systems, not just "high-risk" AI.

Meanwhile, AI guidelines under discussion in the US are a study in contrast compared to the EU's AI Act: they are voluntary, nonprescriptive, and focused on changing the culture of tech companies. The White House Office of Science and Technology Policy released a blueprint for an AI Bill of Rights in October 2022, proposing a set of five principles and practices: (i) safe and effective systems; (ii) algorithmic discrimination protections; (iii) data privacy; (iv) notice and explanation; and (v) human alternatives, consideration, and fallback.[258] In the wake of the release of ChatGPT, the White House has announced its support for public assessments of generative AI systems to educate the public about whether these new systems align with the administration's AI Bill of Rights that is aimed at preventing misuse. Until the time this paper was prepared, there had been no serious consideration of a US analogy to the EU AI Act or any sweeping federal legislation to govern the use of AI, nor is there any substantial state legislation in force. That said, the National Institute of Standards and Technology (NIST), the Federal Trade Commission, and the Food and Drug Administration have provided recent guidance on how the US government may look to govern AI in the future.

In January 2023, NIST released the AI Risk Management Framework 1.0 (RMF) as a voluntary, non-sector-specific, use-case-agnostic guide for technology companies that are designing, developing, deploying, or using AI systems to help manage the many risks of AI.[259] NIST's RMF aims to incorporate trustworthiness considerations into the design, development, use, and evaluation of AI products, services, and systems. The reasoning is that trustworthy AI systems will help preserve civil liberties and rights and enhance safety, while creating opportunities for innovation and realizing the full potential of this technology. To that end, the RMF encourages organizations to adopt a risk mitigation culture that allocates resources according to the risk level and impact of AI systems, while acknowledging that incidents and failures cannot be eliminated.

257  L. Edwards. 2022. Expert Opinion: Regulating AI in Europe. *Ada Loevelace Institute*. 31 March. https://www.adalovelaceinstitute.org/report/regulating-ai-in-europe/.

258  The White House, Office of Science and Technology Policy. Blueprint for an AI Bill of Rights: Making Automated Systems Work for the American People. https://www.whitehouse.gov/ostp/ai-bill-of-rights/.

259  E. Tabassi. 2023. *Artificial Intelligence Risk Management Framework (AI RMF 1.0)*. Gaithersburg, Maryland: National Institute of Standards and Technology. https://doi.org/10.6028/NIST.AI.100-1.

NIST defines trustworthy AI systems as being (i) valid and reliable; (ii) safe in the sense of not causing physical or psychological harm or endangering human life, health, property, or the environment; (iii) fair with regard equality and equity and free of computational or human bias; (iv) secure and resilient to withstand adversarial attacks or unexpected changes in their environment or use; (v) transparent and accountable; (vi) explainable and interpretable; and (vii) privacy-enhanced, whereby privacy values such as anonymity, confidentiality, and control should guide choices for AI system design, development, and deployment.

The NIST guidelines do not prescribe risk thresholds or risk values. Rather, it is up to developers to weigh the risks and advantages of AI systems. In practice, NIST's guidelines suggest that US tech companies might submit to a lot of outside oversight when they create their AI products. NIST recommends that an independent third party or experts who did not serve as front-line developers weigh the advantages and disadvantages of an AI system, consulting stakeholders and impacted communities. NIST also wants AI developers to ensure they have workforce diversity to make sure AI works for everyone. A major caveat is that AI developers are deciding whether their systems are too risky—or not.

Unlike the EU AI Act, the AI RMF 1.0 by NIST is designed to facilitate the management of risks for any type of AI system, not just those classified as high risk. Representatives from several stakeholder groups are supposed to be involved in the risk analysis, a key feature being given little play in the proposed EU AI Act. The latter relies on a regulatory process that identifies, analyzes, and evaluates pre- and post-deployment risks followed by the adoption of risk management measures. Provided organizations comply with these conditions, they are free to generate their own risk management systems. NISTS's RMF, on the other hand, relies on the alignment of incentives; in other words, an entity's self-interest in the adoption of this framework. For instance, entities can gain reputational goodwill from customers and protection in judicial proceedings by using best practices as mitigating factors; they can also synchronize the practices of supply chain providers, for example. Furthermore, NIST created a structure to guide entities into managing the risks of their own AI systems.[260]

What lessons could governments wanting to genuinely promote democratic values draw from these three approaches? [261]

First, many of the consensus AI principles that are often cited as supporting democratic values can just as easily be used for authoritarian purposes. Ensuring that AI systems serve broader democratic goals requires not only technical scrutiny but also strong institutions and oversight mechanisms that allow the public to understand how regulators apply discretion in enforcement. Demonstrating that AI applications live up to principles such as "fair," "accountable," and "human-centered" will require resources and inclusive processes to collect input from diverse publics about what they want and what they do not want from new technologies.

[260] M. Brakel. 2022. Lessons from US NIST for EU AI Act. *European Commission Working Group 1-Technology Standards*. 28 June. https://futurium.ec.europa.eu/en/EU-US-TTC/wg1/documents/lessons-us-nist-eu-ai-act-future-life-institute.

[261] M. O'Shaughnessy. 2023. What a Chinese Regulation Proposal Reveals About AI and Democratic Values. *Carnegie Endowment for International Peace*. 16 May. https://carnegieendowment.org/2023/05/16/what-chinese-regulation-proposal-reveals-about-ai-and-democratic-values-pub-89766.

Second, many of the most consequential uses of AI systems, from the provision of critical services to policing and surveillance, are situated within government agencies. Regulations that apply only to the private sector ignore some of the strongest potential negative impacts on democratic principles. Recent efforts to increase oversight of government AI use and to institute guardrails applicable to high-risk government AI uses are welcome developments. But debates over how rules for AI should apply to law enforcement or how to limit adversarial dual use demonstrate that these issues are rarely clear cut.

Third, an embrace of AI systems that erode individual privacy, perpetuate discrimination, and widen inequality are more likely to undermine public trust in industry and government rather than a furthering of geostrategic advantages.

# 7.2. Assessing Ethical AI Risks at an Organizational Level

Many organizations are facing practical challenges from a risk management perspective in navigating these issues and guiding the development of ethical AI (eAI) systems. Effective and continuous risk management needs to be a central component of ethics and compliance programs to anticipate and mitigate eAI risks before they occur.

Available resources, including standards (e.g., ISO), guidelines, self assessments, ethics-based auditing, and a handful of tools that address individual challenges, have come under criticism. They tend to be too abstract or technology centered, do not focus on human rights, are disconnected from organizational realities, lack validation mechanisms, and/or are incompatible with standardized risk assessment models. Yet, new laws and regulations, such as the proposal by the European Commission for an AI Act or Germany's forthcoming Act on Corporate Due Diligence in Supply Chains, will raise the bar by introducing new demands on assessments to manage and mitigate high-risk use of AI systems.

The challenge will be to develop innovative risk management approaches for managing AI to assure the legality, ethics, and technical robustness of AI systems and detect negative AI externalities that, left unaddressed, could otherwise infringe on legal and human rights. Considering the rapidly evolving eAI landscape, such a risk assessment methodology must be sufficiently adaptable to apply to a wide variety of contexts.

What follows is a summary of a data-driven risk assessment methodology for eAI[262] that has been developed and field tested in approximately 200 use cases by a team from Sweden's Lund University on the Ethical AI Governance Platform[263] for use by organizations (Figure 25).[264]

---

[262] A. Felländer et al. 2022. Achieving a Data-Driven Risk Assessment Methodology for Ethical AI. *Digital Society*. 1 (2). 1–27. https://doi.org/10.1007/s44206-022-00016-0.

[263] anch.AI. https://anch.ai/.

[264] A comparison of this model with other frameworks, such as the newly released IEEE Standard Model for Addressing Ethical Concerns during System Design (IEEE 7000-2021) or Ernst & Young's recent survey of AI risk assessment methodologies, indicates strong parallels. Notwithstanding some differences in formulation, these approaches are similar in focusing on categorization of risk, requirements for trustworthiness of AI, transparent communications, traceability of ethical values, and software engineering methods to help address ethical concerns during system design. See S. Spiekermann. 2022. What to Expect From IEEE 7000: The First Standard for Building Ethical Systems. *IEEE Technology and Society*. 21 February. http://technologyandsociety.org/what-to-expect-from-ieee-7000-the-first-standard-for-building-ethical-systems/; and G. Ezeani et al. 2022. *A survey of artificial intelligence risk assessment methodologies: The global state of play and leading practices identified*. London: Ernst & Young and Trilteral Research. https://www.trilateralresearch.com/wp-content/uploads/2022/01/A-survey-of-AI-Risk-Assessment-Methodologies-full-report.pdf.

To establish the requirements for a standardized risk assessment methodology, a multi-stakeholder group first identified the following most common and distinctive risks:[265]

- **Privacy intrusion,** whereby AI and data-driven solutions interfere with personal or sensitive data without regard to consent of the individual or groups whose data is collected, how data is shared or stored, compliance with the law, or other legitimate needs to protect the best interests of an individual or groups (Right to privacy).
- **Amplified discrimination,** whereby AI and data-driven solutions cause, maintain, or increase prejudicial decisions, treatment, and/or biases toward race, sex, or any other protected groups obliged to equal and fair treatment (Right to fair treatment).
- **Violation of autonomy and independent decision-making,** whereby AI and data-driven solutions, whether intentionally or not and without consent, facilitate behavioral changes that manipulate independent decision-making and social well-being (Right to autonomy).
- **Social exclusion and segregation,** whereby AI and data-driven solutions contribute to an unfair denial of resources, rights, goods, and the ability to participate in normal relationships and activities, whether in economic, social, cultural, organizational, or political arenas (Right to inclusion).
- **Prevention of access to public service,** whereby AI and data-driven applications contribute to a denial of public social assistance and service (Right to public services access).
- **Harm to safety,** whereby AI and data-driven applications enable unwanted physical harm to an individual, group, or organization stemming from underdeveloped AI system that is attributed to negligence from an organization (Right to physical safety and security).
- **Harm to security of information,** whereby AI and data-driven solutions facilitate potential damage from unauthorized access to private data, due to faulty data protection and processing, or criminal activity (Right to security of information).
- **Misinformation and disinformation,** whereby AI and data-driven solutions, whether intentionally or not, distribute information that is regarded as false and harmful to society (Right to be informed).

Based on this eAI landscape review, the team identified four root causes of risks (also known as *pitfalls*):

- **Misuse or overuse of data,** whereby the AI application could be overly intrusive, use private data, or be used for unintended purposes by others.
- **Bias of the creator,** whereby values and biases are intentionally or unintentionally programmed by the developer, who may also lack knowledge and/or skills of how the solution could scale.
- **Immature data and AI,** whereby insufficient training of algorithms on data sets as well as lack of representative data could lead to incorrect and unethical recommendations.
- **Data bias,** whereby the data available is not an accurate reflection of reality or the preferred reality and may lead to incorrect and unethical recommendations.

---

[265] The rapid spread of AI systems has precipitated a rise in ethical and human rights-based frameworks intended to guide the development and use of these technologies. To help understand and map this expanding universe of principles, Harvard's Berkman Klein Center has compared and visualized the contents of 36 prominent AI principles documents side by side. This effort uncovered a growing consensus around eight key thematic trends: privacy, accountability, safety and security, transparency and explainability, fairness and nondiscrimination, human control of technology, professional responsibility, and promotion of human values. These principles mirror the AI risks, pitfalls, and fundamentals put forward by the team at Lund University. J. Fjeld et al. 2020. *Principled Artificial Intelligence: Mapping Consensus in Ethical and Rights-Based Approaches to Principles for AI*. Boston: Harvard University Berkman Klein Center for Internet & Society. https://dash.harvard.edu/handle/1/42160420.

**Figure 25:** Overview of Data-Driven Risk Assessment for Ethical AI

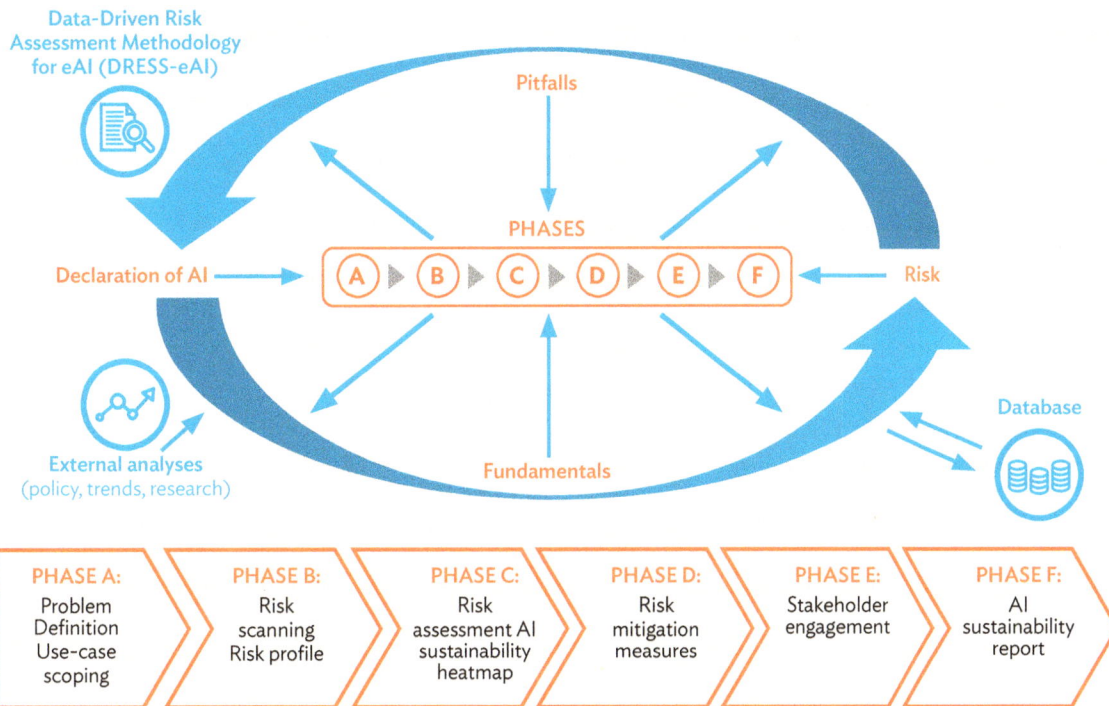

AI = artificial intelligence, eAI = ethical AI.
Source: Felländer et al. 2022. Achieving a Data-Driven Risk Assessment Methodology for Ethical AI. *Digital Society*. 1 (13).

To address these pitfalls, any organization aiming to achieve eAI needs to put certain structural foundations (also known as *fundamentals*) in place as a minimum requirement. These include the following:

- The need to establish accountability and justify one's decisions and actions to partners, users, and others with whom the AI system interacts.
- The adoption of governance principles, policies, and/or protocols, as well as continuous monitoring of their proper implementation.
- The ability to explain algorithmic or data-driven decisions such that they can be understood by end users and other stakeholders using nontechnical terms to establish trust.
- The requirement for transparency to discover, trace, and detect how and why an AI system made a specific decision or acted in a certain way, and, if a system causes harm, to discover the root cause.

Next, the proposal outlines a structured process to ensure that human values and rights are sustained for data-driven AI applications. The six process phases are modeled based on the widely used ISO standard 31000:2009 for risk management,[266] with each phase being identified as a necessary step for systematically ensuring rigorous eAI practices.[267] The methodology reflects the AI systems' life cycle[268] and the recent Declaration of eAI by the European Commission,[269] which will require these types of risk assessments going forward.

To address the frequent lack of cross-functional teams tackling eAI, the suggestion is for all roles and functions (technical, legal, risk, compliance, communications, sustainability, and human resources) to be part of the process. As illustrated in Figure 25, the six phases are structured as follows:

- **Phase A:** Workshops are organized to come up with a detailed description of the eAI use case, including a summary of challenges; guiding policies, codes, and values; key stakeholders; technical specifications; and the project team. This phase also includes a survey to assess the organization's preparedness for ethically high-risk AI systems.
- **Phase B:** Using an in-depth questionnaire, the risk scanning and profiling helps to generate structured data related to the organization's current state of achieving *fundamentals* and identifying vulnerabilities associated with *pitfalls*. Figure 26 demonstrates the mapping of *pitfalls* along the AI life cycle. This risk survey can also be used as a tool to identify possible gaps between stated ethical principles and what might be happening at product or organizational level.
- **Phase C:** The risk assessment phase identifies and prioritizes ethical risk scenarios as well as the stakeholders which could be impacted by their exposure to *pitfalls* and *fundamentals*.
- **Phase D:** In this phase, technical and/or nontechnical risk mitigation measures are determined at both organizational and use-case levels to address the root causes of a risk scenario, or its effects. The implementation of risk-mitigating measures (e.g., updated legal documents, synthetic data for avoiding bias or preserving privacy, tailored explainability models, training, establishment of an AI ethics board) is assigned to risk owners within the organization, together with plans for implementation and continued monitoring of follow-up actions.
- **Phase E:** A critical validation step is to capture stakeholder feedback on the risk mitigation measures and what should be done to manage actual and potential impacts.
- **Phase F:** A subsequent review, including an updated risk scan of the use case, is conducted to track the effectiveness of risk mitigation activities and provide recommendations on how internal frameworks can be strengthened.

266  ISO. ISO 31000:2009(En), Risk Management—Principles and Guidelines. https://www.iso.org/obp/ui/#iso:std:iso:31000:ed-1:v1:en.
267  European Commission. A European Approach to Artificial Intelligence. https://digital-strategy.ec.europa.eu/en/policies/european-approach-artificial-intelligence.
268  S. Flynn. 2022. Machine Learning Life Cycle: 6 Stages Explained. *CIO Insight*. 22 July. https://www.cioinsight.com/big-data/machine-learning-life-cycle/.; and M. Haakman et al. 2021. AI Lifecycle Models Need to Be Revised. *Empirical Software Engineering*. 26 (5): 95. https://doi.org/10.1007/s10664-021-09993-1.
269  European Commission. Declaration on European Digital Rights and Principles. https://digital-strategy.ec.europa.eu/en/library/declaration-european-digital-rights-and-principles.; and Think Tank European Parliament. European Declaration on Digital Rights and Principles. https://www.europarl.europa.eu/thinktank/en/document/EPRS_BRI(2022)733518.

**Figure 26: Mapping Ethical AI Pitfalls along the AI Life Cycle**

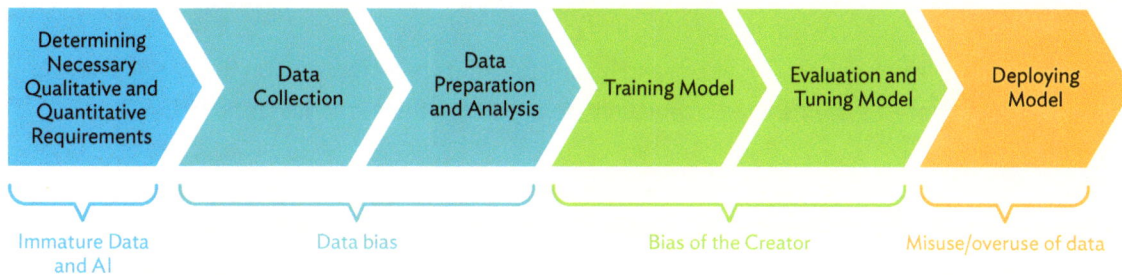

AI = artificial intelligence.
Source: Felländer et al. 2022. Achieving a Data-Driven Risk Assessment Methodology for Ethical AI. *Digital Society*. 1 (13).

This eAI risk assessment approach has been applied to a variety of real-world situations:

- In one use case, the AI model was used to classify job seekers based on personal data related to job hiring, education, language proficiency, as well as labor market conditions. The intended outcome of the AI model was a prediction of how far from the labor market a job seeker is, placing him or her in one of three categories based on a rule-based selection. After going through Phases A, B, and C, it became obvious that this use case had several ethical and societal *pitfalls* ("Bias of the creator" and "Data bias") and revealed weaknesses in *fundamentals* ("Accountability" and "AI governance"). Nine risk scenarios were likely to occur with severe impacts on people and society. Mitigating measures were identified in Phase D for each risk scenario, including updating the user interface to prevent misuse of model outputs, clarifying the purpose statement for the various stakeholders, and implementing methods for explaining model outputs. With these modifications in place, the solution could be scaled.

- In a second use case, the AI system was intended to monitor and select transactions on third-party platforms that should be reviewed for potential tax fraud. By performing the eAI risk assessment at the planning stage, the Phase A review highlighted the absence of a systematic approach and accountability mechanisms to detect and handle eAI risks, an inability to monitor the AI system, a large exposure to "Data bias" and "Bias of the creator,"; and competency gaps. After taking mitigating actions over the course of the next year, the organization streamlined the implementation, clarified transparency and explainability, established an eAI policy, and is considering an AI ethics committee.

This eAI risk assessment framework does however have limitations: first, while the model is intended for organizations that are actively seeking to apply AI, it does not support quantitative assessments of algorithmic bias and fairness (for examples, see following section); second, this framework does not provide a full compliance assessment tailored to regulations; third, with its emphasis on cross-functional participation and questionnaire-based data collection, its playbook foregoes specific checklists or guidelines in favor of a step-by-step approach to assess, report, and monitor comprehensive eAI risk exposure and mitigation plans at both organizational and use-case levels; and fourth, given demands for attaining AI expertise from IT, business, and legal perspectives, this framework can most realistically be applied to organizations with at least a basic level of AI maturity.

# 7.3. AI Ethics Tool Kits

Technology teams, researchers, policymakers, and business leaders have focused on the design and development process of AI systems as an opportunity to promote ethical and fair approaches.

Tool kits play an important role in this endeavor as they can help technology practitioners and other stakeholders address ethical issues and condense a very complex and technical debate into practical guidance. AI ethics tool kits operate on the belief that ethical issues which may arise in AI design can be addressed with the right tools or processes. Clearly, the existence of tool kits is insufficient without adoption, used in practice, and adaptation to a wide range of contexts within organizations, collaborative ventures, and broader society.

To support AI practitioners in addressing the ethical dimensions of AI, resources include open-source or proprietary code, scenario simulations, checklists, life-cycle strategies, risk matrices, and reflection exercises. As the field of AI ethics has moved from developing high-level principles to operationalizing them in a set of practices, the central question is how these practices can be adapted and implemented in the social and organizational context where AI systems are conceived, piloted, deployed, evaluated, and revised. Technical decisions on machine learning models are rarely ever just technical in nature and are often contested and negotiated among multiple parties (e.g., data scientists, business team members, product managers). A key question is to what extent such organizational dynamics and trade-offs shape AI ethics work in practice.

For this paper, a cross-section of tool kits was selected based on three questions:

- Who is the intended audience?
- Does the tool kit provide specific guidance or recommended actions?
- Are the tool kits being used?

Table 15 offers a classification of AI ethics tool kits based on author(s), intended audience, stated goal(s), and reference material. Given recent developments in AI technologies, the available materials have increased dramatically in number. Users of this primer should consider recent tool kits published after the advent of widely available generative AI in late 2022, which have not been covered here.

Authors include international organizations, governments, technology companies, academic institutions, think tanks, nonprofit organizations, open-source communities, design agencies, and individual tech workers.

Target audiences include technical teams (data scientists, designers, product teams); different functions and hierarchies within organizations (team members, managers, executive leadership, board of directors); and a range of external stakeholders (policymakers, government leaders, advocates, software customers, civil society organizations, community groups, and users).

**Table 15:** Overview of AI Ethics Tool Kits
(Covering a Sample from 2018 to 2022)

| | Tool Kit Name | Tool Kit Author | Author Types | Audiences | Format |
|---|---|---|---|---|---|
| **1.** | **International Standards and Recommendations** | | | | |
| T1 | Ethically Aligned Design (Ver. 2)<br><br>AI in Business | Institute of Electrical and Electronics Engineers (IEEE) | International Network of Standard Setters | Global AI Community, Standard Setters, Business, | PDF Guides, Principles, Dual Use Technologies, AI Readiness Framework, Standards and Certification Programs, Metrics and Methodologies, Reference Material |
| T2 | A Decision Maker's Toolkit of AI (2021) | UNESCO | International Organization | Global science and education community | Consensus Recommendations (on Tech Foresight, Implementation Guides, Model Use Cases, Policy Process, Online Resources) |
| T3 | Policy Guidance on AI for Children (version 2.0) (2021) | UNICEF | International Organization | Governments, Private Sector, Parents, Teens | PDF Guide, Roadmap for Policymakers, Guides for Parents and Teens, Case Studies |
| T4 | Ethics and Governance of AI for Health (2021) | World Health Organization (WHO) | International Organization | AI Tech Developers, Ministries of Health, Health Care Providers, Gov't Agencies | PDF Guide, Legal and Policy Provisions for Health, Liability and Governance Regimes |
| T5 | International Standards to Enable Global Coordination in AI Research and Development (2019) | Future of Humanity Institute and Oxford University | Think Tank, University | International AI Community, Standard Setters | PDF Guide, AI Standards |
| **2.** | **Tool Kits for Governments** | | | | |
| T6 | Model AI Governance Framework (2nd ed) (2020) | Singapore's Infocomm Media Authority and Personal Data Protection Commission | Government Agency | Business, AI Solution Providers, Customers, International AI Community | PDF Guide, Umbrella Framework, Self-Assessment Worksheets, Risk Matrix, Use Cases, |
| T7 | Ethics, Transparency and Accountability Framework for Automated Decision-Making (2021) | UK Cabinet Office | Government Agency | Government Departments | Umbrella Framework, Reference Documents, Case Studies, |

*continued on next page*

**Table 15** *continued*

| | Tool Kit Name | Tool Kit Author | Author Types | Audiences | Format |
|---|---|---|---|---|---|
| **T8** | AI and Data Protection Risk Toolkit (2021) | UK Information Commissioner's Office (ICO) | Government Agency | Compliance Officers, General Counsel, Risk Managers, Tech Teams | Self-Assessment Worksheets, Risk Matrix, Documentation/Reference Material, |
| **T9** | Data Ethics Workbook (2018) | UK Central Digital and Data Office | Government Agency | Data Practitioners, Policymakers | PDF Guide, Self-Assessment Questionnaire, Case Examples |
| **T10** | Understanding AI and Safety in the Public Sector (2019) | The Alan Turing Institute | National Institute | Public Sector | PDF Guide, AI Life Cycle, Documentation/Reference Material |
| **T11** | Towards Responsible AI for All (2021) | India's Niti Aayog | Government Think Tank | Government, Private Sector, Academia | PDF Guides, Self-Assessment, Legal and Regulatory Approaches, Operational Examples |
| **T12** | Guidance for Regulation of Artificial Intelligence Applications (2020) | US Government | Government Agency | Public Sector Agencies | Umbrella Framework, Regulatory and Non-regulatory Provisions |
| **3.** | **Tool Kits for Cities** | | | | |
| **T13** | AI System Ethics Self-Assessment Tool (2020) | Digital Dubai Authority | Government Agency | City Ecosystem, Industry, Academia | Self-Assessment Guide (download of worksheet subject to registration) |
| **T14** | Ethics and Algorithms Toolkit (2018) | JHU's Centers of Gov't Excellence and Civic Impact, City of San Francisco, Harvard DataSmart, Data Community DC | University, Government Agency, Non-Profit | Gov't Leaders, Stakeholders, Data Analysts, IT-experts, AI-vendors | Guide, Worksheet |
| **4.** | **Business** | | | | |
| **T15** | Assessment Methodologies for Financial Institutions (2022) | Monetary Authorities of Singapore and Veritas Consortium | Financial Regulator and Industry Consortium | Financial Institutions | PDF Guide, Checklist, Assessment Methodologies, Use Cases, Open Source Code |
| **T16** | Aletheia (2021) | Rolls Royce | Business | Cross-industry and society, Product Teams, Risk Managers | AI Life Cycle, Risk Assessment Template, Case Studies, |
| **T17** | Empowering AI Leadership (2019) | World Economic Forum | International Nongovernment Organization and Lobbying Organization | Board of Directors | PDF Modules (Strategy, Control, Oversight), Agenda-Setting Guides, Examples, |
| **T18** | Responsible AI (2021) | PWC | Consulting Company | Business | Diagnostic Survey, Frameworks, Examples |

*continued on next page*

**Table 15** *continued*

| | Tool Kit Name | Tool Kit Author | Author Types | Audiences | Format |
|---|---|---|---|---|---|
| **T19** | Foresight into AI Ethics (FAEI) Toolkit (2021) | Open Roboethics Institute (ORI) | Think Tank | Data Scientists, Engineers, Product Teams, Business Leaders, | Foresight, Stakeholder Maps, Use Cases, Robotics |
| **5.** | **Civil Society Engagement** | | | | |
| **T20** | Digital Impact Toolkit (2022) | Digital Civil Society Lab @ Stanford | University | Civil Society Organizations | Checklists, Worksheets, Reading Materials |
| **T21** | Algorithmic Equity Toolkit (2020) | ACLU of Washington, Critical Platform Studies Group, Tech Fairness Coalition | University, Nonprofit | Community Groups | Activities |
| **T22** | Algorithmic Accountability Policy Toolkit (2018) | AI Now Institute | University Research Center | Legal and Policy Advocates | PDF Guide, Advocacy Tools, Case Studies |
| **T23** | From Principles to Practice: An Interdisciplinary Framework to Operationalize AI Ethics (2019) | AI Ethics Impact (AIEI) Group | Consortium (Technical-Scientific Association, Foundation, Universities) | Policymakers, Regulators, Oversight Bodies, Standard Setters, Technology Teams | PDF Guide, Risk Matrix, Case Studies |
| **T24** | Dynamics of AI Principles (2020) | AI Ethics Lab | Consulting and Research | Government, Private Sector, Academics | Global mapping and comparison of AI Principles, Documentation, Guidebook |
| **6.** | **Humanitarian Assistance** | | | | |
| **T25** | AI Ethics Toolkit (2020) | Data Science and Ethics Group | Network of data scientists, researchers, humanitarians, ethicists | Humanitarian Organizations | Reference Material, Decision Tree, Case Studies, |
| **T26** | AI Ethics for Non-Profits Toolkit (2020) | NETHOPE Solutions Center (w/ US AID, MIT D-Lab, Plan Int'l, Catholic Relief Services, Humanitarian OpenStreet Maps Team) | Network of digital nonprofits and development organizations | Nonprofits, Development Agencies, Humanitarian Organizations | Facilitators Guide, Workshop Deck, Case Studies |
| **7.** | **Activity-Based Tool Kits** | | | | |
| **T27** | Responsible Innovation Toolkit (Judgment Call, Harms Modeling, Community Jury) (2021) | Microsoft | Technology Company | Product Teams, Technology Builders | Activity |
| **T28** | HAX Toolkit (2020) | Microsoft | Technology Company | UX, AI, Project Management and Engineering Teams | Guide, Workbook/ Worksheets, Examples, Guidelines, Activity |

*continued on next page*

**Table 15** *continued*

| | Tool Kit Name | Tool Kit Author | Author Types | Audiences | Format |
|---|---|---|---|---|---|
| **T29** | Ethical Toolkit for Engineering and Design Practice (2018) | Markkula Center for Applied Ethics @ Santa Clara University | University | Product Teams, Technology Builders | Activity, Teaching Guide |
| **T30** | Ethical OS Toolkit (2018) | Institute for the Future, Omidyar's Tech and Society Solution Lab | Think Tank | Technology Teams, Product Managers, Engineers | PDF Guide, Scenarios, Risk Matrix, Use Cases |
| **8.** | **Tool Kits for Tech Teams and Data Scientists** | | | | |
| **T31** | Audit-AI (2020) | Pymetrics | Technology Company | Data Scientists | Open Source Code, Documentation, Examples |
| **T32** | What-If Tool (2021) | Google (AI Research Team) | Technology Company | Data Scientists | Open Source Code, Tutorials, Documentation, Examples |
| **T33** | TensorFlow Fairness Indicators (2022) | Google | Technology Company | Data Scientists, Product Teams | Open Source Code, Documentation, Examples |
| **T34** | AI-Explainability AI-Fairness 360 (2020) | IBM | Technology Company | Data Scientists | Open Source Code, Documentation, Code Examples, Tutorials |
| **T35** | InterpretML (2021) | Microsoft | Technology Company | Data Scientists | Open Source Code, Documentation, Code Examples |
| **T36** | FairLearn (2021) | Microsoft Research, Open Source Community | Technology Company, Open Source Community | Data Scientists | Open Source Code, Documentation, User Guide, Code Examples |
| **T37** | Deon Ethics Checklist (2020) | Driven Data Lab | Non-Profit | Developers | Checklist, Open Source Code, Documentation |
| **T38** | Design Ethically Toolkit (2022) | Designers/ Engineers from various Tech Firms | Tech Workers | Designers | Exercises, Worksheet |
| **T39** | SageMaker Clarify (2021) | Amazon | Technology Company | AWS Customers | Proprietary Code, Documentation, Example |
| **T40** | Aequitas (2018) | Center for Data Science and Public Policy | University | ML Developers, Analysts, Policymakers | Open Source Code, Web Audit Tool, Example, Documentation |
| **T41** | Handbook on Non-Discrimination by Design (2022) | Tilburg Institute for Law, Technology and Society | University | ML Developers, Legal and Rights Advocates, Stakeholders | PDF Guide, Questionnaire and Discovery Tools, Case Examples, |

ACLU = American Civil Liberties Union; AI = Artificial Intelligence; PDF = portable document format; PWC = PricewaterhouseCoopers International Limited; UNESCO = United Nations Educational, Scientific and Cultural Organization; UNICEF = United Nations International Children's Emergency Fund; UK = United Kingdom; US = United States; UX = user experience.
Source: Author's compilation.

Following are ethical problems that tool kits address:

- Identifying how AI systems can have harmful effects on people, for instance by leading to unfair outcomes or contributing to unintended consequences. Examples for mitigating algorithmic biases come from both the private sector, such as IBM's open-source software tool kit AI Fairness 360 or Amazon's SageMaker Clarify, and draft laws by government agencies and Parliaments, such as those of the EU, the UK, and India.
- Anticipating potentially harmful impacts on the organization that is developing or deploying the AI systems. These tool kits relate AI ethics to potential business, financial, or reputational risks, or corporate risk management more broadly. To help organizations prepare contingencies for possible negative outcomes, their advice is to establish governance mechanisms or generate compliance reports to manage these risks.
- Offering advice on AI ethics in the pursuit of positive outcomes, providing business leaders and risk professionals with arguments and illustrations why fairness and transparency are good for organizations.

Some tool kits take the view that the positive or negative impacts of AI technologies will reverberate at a global scale, helped by the rapid deployment of data systems and algorithms. Designing AI systems to be trustworthy requires solutions that are based on ethical principles, deeply rooted in widely shared and timeless values. Framing AI ethics on a global scale draws attention to the diversity of societies that interact with AI systems and assumes that it is possible to converge on a broadly shared universal definition of social values and their impacts. This is an ambitious goal which requires international collaboration to forge a consensus (e.g., UNICEF's 2-year effort to build global support for its "Guidance on AI for Children").[270]

When it comes to ethical problems and the question of how to address them, tool kits claim legitimacy by linking up to existing frameworks or widely accepted practices. Several tool kits frame AI ethics as upholding and promoting the UN's Guiding Principles on Business and Human Rights.[271] Other tool kits refer to compliance with non AI-specific laws and standards as a legitimate basis for ethical action.[272] Some tools, for instance, measure discriminatory patterns in data and machine-learning predictions by comparing the results with regulatory standards.[273] Yet, other tool kit authors align themselves with the notion of responsible innovation,[274] without, however, clarifying what this means in practice.[275]

[270] UNICEF. AI for Children. https://www.unicef.org/globalinsight/featured-projects/ai-children.
[271] Business & Human Rights Resource Centre. UN Guiding Principles on Business and Human Rights. https://www.business-humanrights.org/en/big-issues/un-guiding-principles-on-business-human-rights/.
[272] A. Ritchie and S. Clarke. 2019. The ethics of artificial intelligence: laws from around the world. *MinterEllison*. 3 June. https://www.minterellison.com/articles/ethics-of-artificial-intelligence-laws-around-the-world.
[273] US Equal Employment Opportunity Commission. 2021. EEOC Launches Initiative on Artificial Intelligence and Algorithmic Fairness. Press release. 28 October. https://www.eeoc.gov/newsroom/eeoc-launches-initiative-artificial-intelligence-and-algorithmic-fairness.
[274] J. Hankins. 2015. What Does "Responsible Innovation" Mean? *IEEE Spectrum*. 24 June. https://spectrum.ieee.org/what-does-responsible-innovation-mean.
[275] Microsoft Ignite. Responsible Innovation: A Best Practices Toolkit. https://docs.microsoft.com/en-us/azure/architecture/guide/responsible-innovation/.

Tool kits may also target different job functions. For instance, several tool kits focus on specific roles within an organization, such as software engineers, data scientists, risk or internal governance teams, senior executives, or board members. Tool kits targeted at data science or engineering roles tend to regard ethics as meeting certain technical specifications; others see individuals or development teams within the organization as catalysts who, once they become aware of possible ethical issues, are able to change the direction of the design process toward more ethical outcomes. The implicit theory of change relies on having group-level discussions with well-intentioned and well-informed actors, who share a profound awareness of ethical issues. For senior management and board members, tool kits frame ethics as both a business risk and strategic differentiator in a crowded market. What remains unclear is how differences of view, expertise, and authority within teams or between technical staff and management are to be resolved.

Other tool kits seek to include external stakeholders that may be impacted by the technology, including clients, vendors, customers, civil society groups, advocacy groups, or communities (Box 7). These tool kits suggest that concerns should be raised and inputs provided from the ground up. None, however, address how to select stakeholders, or how to engage with them in either understanding the ethical impact of AI systems or involving them in the design process.

Several tool kits portray ethics as an integral part of the technical design of machine learning models and call for an engagement with other business groups or external stakeholders to solicit their inputs about potential ethical harms, or for the AI design team to inform others about ethical risks. However, these tools offer little guidance on how nontechnical stakeholders can influence key design decisions. This leaves open the question whether individual self-reflection or team discussions might lead to meaningful shifts. For instance, some tools are designed to be integrated into specific machine learning programming suites to measure biases that can occur during the machine learning life cycle.

---

**Box 7: Piloting Algorithmic Impact Assessments in the United Kingdom's Health-Care System**

Data-driven technologies have made health data more accessible for research and innovation—and subject to new risks. These include bias (exacerbating existing health inequalities), diminished transparency (inaccessible language makes it hard for clinicians to use and undercuts public trust), and violation of privacy (normalization of surveillance).

A research partnership between the United Kingdom's National Health Services (NHS) and the Ada Lovelace Institute is piloting an algorithmic impact assessment (AIA) in health-care imaging which has mapped out a detailed, step-by-step AIA process in real-world use cases involving both public and private sector. The NHS AI Lab can leverage these research insights to develop its AIA process for data access, helping developers and researchers think through the potential impacts of the technologies developed using NHS data.

Building trust in the use of AI technologies for screening and diagnosis is fundamental if the NHS is to realize the benefits of AI. The new AIA process aims to ensure that AI researchers and companies meaningfully engage with patients and health-care professionals way ahead of bringing an AI solution to market and identify potential risks before getting access to NHS extensive patient data of chest x-rays, MRIs, and other images to test and validate medical AIs.

Sources: Ada Lovelace Institute. Algorithmic Impact Assessment in Healthcare. https://www.adalovelaceinstitute.org/project/algorithmic-impact-assessment-healthcare/.; and NMIP Algorithmic Impact Assessment (AIA) Template. https://docs.google.com/document/d/12HXv7Kb4dZLnA0BkL7DiccBxoq-SIg2meBsUBq_QQQI/edit?usp=embed_facebook.

Finally, some tool kits propose that external pressure (i.e., media, public advocacy, academic institutions, civil society) may lead to changes in AI design and deployment, including by compelling government agencies to provide more transparency and accountability in how their algorithmic systems are being used.

Technology practitioners, business groups, and policymakers rely on tool kits to help guide and support their work of AI ethics.

New approaches are beginning to emerge to move public debate and knowledge exchange beyond tool kits. These include alternative governance arrangements and the participation of policymakers, civil society, or community stakeholders in shaping the design, deployment, oversight, and assurance of AI systems. For instance, more than 2,000 participants from across the globe joined a series of public workshops by the US NIST[276] to solicit feedback on an AI Risk Management Framework.[277]

276 National Institute of Standards and Technology. Building the NIST AI Risk Management Framework: Workshop #2. https://www.nist.gov/news-events/events/2022/03/building-nist-ai-risk-management-framework-workshop-2.
277 E. Tabassi. AI Risk Management Framework: Initial Draft. 17 March 2022.

# 8

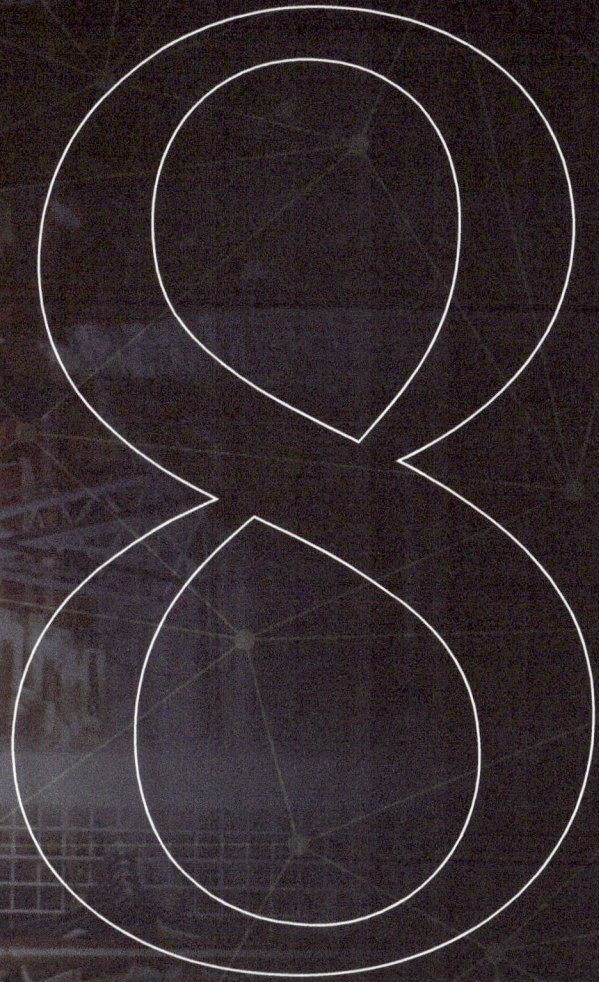

# DIGITAL RISKS TO HUMAN RIGHTS AND VULNERABLE GROUPS

Digital technologies offer new opportunities to uphold or strengthen human rights. All too often, however, digital tools are used to infringe on human rights, be it through data privacy breaches, mishandling of digital identities, surveillance technologies, misinformation and disinformation, or online harassment. This chapter will explore these digital risks, highlight global trends, and discuss policies and practices by governments and major platform companies as well as recent digital governance initiatives at city level.

Another dimension of digital risks to human rights involves vulnerable groups and marginalized communities, including children, displaced people, refugees, and migrants, who lack agency and protection and are routinely exposed to surveillance, data collection, and harmful content.

## 8.1. Digital Risks to Human Rights

Digital risks to human rights may include the following:

- curtailment of freedom of information and expression due to intentional internet shutdowns;
- distortion of free speech through social media or speech-based platforms;
- human rights risk due to abuses of surveillance and biometric technology;
- discriminatory biases in algorithms;
- problems of inaccuracy, discrimination, and lack of agency arising from sharing and/or combining data for individual rating systems (e.g., credit checks, student grading systems; health assessments);
- collective privacy violations (e.g., communities are not aware or would not approve of their personal data being collected, collated, and used);
- data exclusion and poor algorithmic decision-making, which may result in certain population groups being excluded from widespread training data sets (e.g., because of skin color) or social protection and other government systems;
- exclusion bias due to faulty data standards and formats (e.g., binary gender classification);
- gaps in the collection of gender data, and, conversely, discriminatory impacts caused by the overrepresentation of marginalized groups in certain data systems (e.g., criminal justice);
- discrimination, exposure to harm, and function creep from digital ID systems; and
- restrictions of water rights due to excessive consumption by data centers.

The UN's Technology Envoy has drawn up a set of recommendations to protect human rights in the digital era (Table 16), which can serve as a reference point against current realities.

The Office of the United Nations High Commissioner for Human Rights (OHCHR) has observed that MDBs frequently include privacy and data security considerations as part of their risk assessment for digital technology projects, but not (yet) other human rights risk factors associated with the various stages of the data cycle (collection, storage, use, and/or reuse).[278]

---

[278] OHCHR's April 2021 memorandum in the context of ADB's Safeguards Review.

## Table 16: Ensuring the Protection of Digital Human Rights

- Place human rights at the center of regulatory frameworks and legislation on digital tech.
- Provide better guidance on the application of human rights standards.
- Address protection gaps created by evolving technologies.
- Discourage blanket internet shutdowns and generic service blocking and filtering of services.
- Establish human rights-based domestic laws and practices for the protection of data privacy.
- Create company-specific actions to protect privacy rights and other human rights.

- Adopt and enhance safeguards for digital identity protection.
- Protect people from unlawful and unnecessary digital surveillance.
- Adopt laws grounded in human rights to address illegal and harmful online content.
- To ensure safe online spaces, adopt transparent and accountable content governance frameworks that protect freedom of expression, avoid overly restrictive practices, and protect the most vulnerable.
- Ensure a UN system-wide guidance on impact assessment and due diligence in the use of new tech.

UN = United Nations.
Source: UN Office of the Secretary-General's Envoy on Technology. Ensuring the Protection of Human Rights in the Digital Era.

In September 2021, OHCHR published a report titled *The Right to Privacy in the Digital Age.*[279] The report analyzes how AI systems, using profiling, automated decision-making, and machine learning, affect people's fundamental rights and freedoms, including the right to privacy, the right to health, and freedom of expression. Considering AI's ubiquitous effects, OHCHR is calling for a moratorium on the sale and use of AI systems and remote biometric recognition systems in public spaces until adequate safeguards are put in place.

OHCHR is also urging a ban on AI applications that cannot ensure compliance with international human rights law. Concrete steps include the implementation of a robust legislative and regulatory framework, which prevents and/or mitigates any adverse effects of AI on human rights. States are to ensure that any permitted interference with the right to privacy and other human rights by AI does not impair the essence of these rights, that it pursues a legitimate purpose, is necessary and proportionate, and requires adequate justification of AI-supported decisions. OHCHR also advocates that public and private entities systematically conduct human rights due diligence throughout the AI life cycle, increase transparency, and actively combat discrimination.

Freedom House's *Freedom on the Net 2022* report provides a sobering reality check on global internet freedom which declined for the 12th consecutive year, with global practices shifting toward dramatically more state intervention.[280] Among the 70 countries covered in the report, 75% of people live in countries where individuals were arrested for posting content on political, social, or religious issues; 72% live in countries where individuals have been attacked or killed for online activities since 2020; and 64% live in countries where authorities used pro-government commentators to manipulate online discussions.

279  United Nations High Commissioner for Human Rights: The right to privacy in the digital age-Report of the United Nations High Commissioner for Human Rights. https://www.ohchr.org/en/documents/thematic-reports/ahrc4831-right-privacy-digital-age-report-united-nations-high.
280  Human rights online declined in 30 countries and improved in only 18.

At least 20 governments suspended internet access and blocked social media platforms in 2021, often during politically charged periods.[281] The COVID-19 pandemic has allowed states to use smartphone apps for contract tracing and justified targeted surveillance with few safeguards. More than 45 governments are suspected of having obtained private vendors' spyware. In a hopeful sign, the 2022 report noted that a record 26 countries saw improvements in internet freedom. Civil society organizations drove collaborative efforts to improve legislation, develop media resilience, and ensure accountability among technology companies. Successful collective actions against internet shutdowns offered a model for further progress on other problems like commercial spyware.

The private sector is being enlisted to assist government efforts to control online content (e.g., by removing users' speech, eliminating the need for court orders); personal data (e.g., by undermining encryption, using vague exemptions on national security grounds, and onerous licensing requirements); and market behavior (e.g., by imposing large fines for abusive or anticompetitive practices). Thirty-eight countries passed laws affecting the management of user data, with many governments bypassing jurisdictional barriers by requiring companies to store data on local servers, surrender data to law enforcement, or decrypt data when requested.[282] These efforts coincide with growing public anger at data breaches, which further adds pressure to limit pervasive use of personal data by third parties.

At the same time, governments are increasingly clashing with tech companies over the rights of users, forcing companies to comply with censorship and surveillance initiatives. In 2021 alone, at least 21 countries have proposed action to protect competition. However, without proper oversight, these policies can be wielded in a politically motivated manner. Some 48 governments are pursuing administrative or legal action against tech companies, alleging that they used overly broad requirements to moderate and monitor users without safeguards like transparency, judicial oversight, and public accountability. At least 24 countries passed or announced laws governing how platforms are to treat content, which would put freedom of expression at risk, including removal of illegal content or required appointments of a legal representative to manage data at governments' requests. The EU legal framework (i.e., Digital Services Act and the Digital Markets Act) tries to strike a balance by requiring tech companies to report on their practices for content moderation, AI curation, and online ads; and notify users of moderation decisions.[283]

Freedom House's *Freedom on the Net 2021* report makes recommendations to three stakeholder groups: (i) policymakers should protect privacy and security, guarantee competition, transparency, and accountability, foster a reliable and diverse information space, and protect global internet freedom; (ii) companies need to ensure fair and transparent content moderation, resist government orders to shut down or ban digital services, adhere to UN principles and conduct human rights impact assessments, and engage in dialogue with civil society organizations; and (iii) civil society needs to raise awareness about censorship, surveillance, and content manipulation, utilize strategic litigation to push back against shutdowns, and work with policymakers and the private sector to design solutions adapted to the local context.

---

281  For an analysis on the use of AI to generate hyper-targeted disinformation campaigns, data manipulation, and cyber- and AI-enabled cognitive-emotional conflicts during election campaigns in Africa, see E. Pauwels. 2020. The Anatomy of Information Disorders in Africa. Konrad Adenauer Foundation Office New York. 8 September. https://www.kas.de/en/web/newyork/single-title/-/content/the-anatomy-of-information-disorders-in-africa.

282  Freedom House. *Freedom on the Net 2021: The Global Drive to Control Big Tech*. https://freedomhouse.org/report/freedom-net/2021/global-drive-control-big-tech.

283  European Commission. The Digital Services Act package. https://digital-strategy.ec.europa.eu/en/policies/digital-services-act-package.

The Ranking Digital Rights benchmarking initiative[284] scores policies and practices of the 14 largest digital platforms regarding human rights, freedom of expression, and privacy. The goal of this initiative is to create global standards and incentives for platform companies to protect users' rights and improve their policies and practices over time.

The sixth edition of the Big Tech Scorecard was published in 2022 (Figure 27).[285] On the positive side, several companies strengthened their corporate governance practices and established upper-level management oversight on freedom of expression and privacy. The top spot which was occupied by Twitter (now renamed to X) reflected its detailed content policies and release of public data about moderation of user-generated content —policies that have been dismantled following its acquisition by Elon Musk. Microsoft published data on its content restrictions for the first time, while Google's score declined due to outdated policies on notifying users of content restrictions. Meta announced a new human rights policy, but did not commit to upholding international human rights in developing its algorithmic systems.[286] Chinese platforms, historically among the least transparent, improved in response to sweeping regulatory changes. To comply with the PRC's new Personal Information Protection Law, technology firms, such as Baidu, Tencent, and Alibaba are required to give users the option to opt out of algorithmic recommendation and targeting systems. They are also forbidden from unreasonable differential treatment of individuals and price discrimination through automated decision-making.[287]

**Figure 27:** The 2022 Big Tech Scorecard

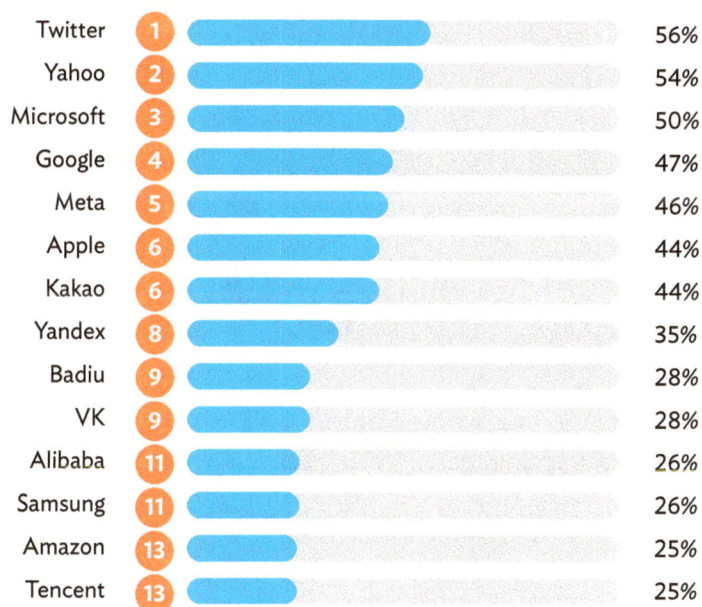

| Rank | Company | Score |
|---|---|---|
| 1 | Twitter | 56% |
| 2 | Yahoo | 54% |
| 3 | Microsoft | 50% |
| 4 | Google | 47% |
| 5 | Meta | 46% |
| 6 | Apple | 44% |
| 6 | Kakao | 44% |
| 8 | Yandex | 35% |
| 9 | Badiu | 28% |
| 9 | VK | 28% |
| 11 | Alibaba | 26% |
| 11 | Samsung | 26% |
| 13 | Amazon | 25% |
| 13 | Tencent | 25% |

Source: Ranking Digital Rights Program, New America Foundation, 2022. https://rankingdigitalrights.org/bts22/.

284  Ranking Digital Rights. Who We Are. https://rankingdigitalrights.org/who-we-are/.
285  Ranking Digital Rights. The 2022 Ranking Digital Rights Big Tech Scorecard. https://rankingdigitalrights.org/index2022/. The rankings do not yet reflect recent decisions by several platform companies (e.g., Twitter [now renamed to X], Microsoft, Meta, Google) to disband their in-house ethics teams and/or sharply cut back on content moderation.
286  M. Sissons. 2021. Our Commitment to Human Rights. *Meta*. 16 March. https://about.fb.com/news/2021/03/our-commitment-to-human-rights/.
287  A. Lee et al. 2021. Seven Major Changes in China's Finalized Personal Information Protection Law. *Stanford University DigiChina Project*. 15 September. https://digichina.stanford.edu/work/seven-major-changes-in-chinas-finalized-personal-information-protection-law/.

As in previous years, every platform again received a "failing" score in the 2022 ranking. Major tech companies and social media platforms have since laid off their content moderation, ethics, and society teams, while at the same time accelerating their push into AI products.[288] Amid increased public scrutiny, Big Tech companies have not disclosed adequate information about how they conduct human rights due diligence,[289] test and deploy algorithmic systems, and use personal data. Companies fared worst on algorithmic transparency. No company has announced a comprehensive human rights impact assessment of algorithms used to target ads to users. There is little evidence that risks regarding targeted advertising are being identified and mitigated. Yet, the social harm widely associated with digital platforms—hate speech, disinformation, election interference—are closely linked to the surveillance advertising business model.

According to Ranking Digital Rights (RDR), achieving a global internet that supports and sustains human rights is a collective effort which will require (i) policymakers to pursue a ban on surveillance advertising, which favors privacy violations and discrimination by algorithm and shift to an approach that respects human rights; and (ii) companies to have transparent, well-enforced policies about ad content, where ads

---

**○ Box 8: Taking Action against the Threats of Misinformation and Disinformation ○**

Algorithmic manipulation, bias, hate speech, misleading information, deepfakes, and other acts have created a global crisis in the information environment crisis that may pose serious threats to humanity. The cost is billions of dollars; millions of lives; and an erosion of trust in science, our institutions, and each other.

The algorithms operating on digital platforms, particularly social media, amplify false or misleading content, making it easier for conspiracy theories and false claims to go viral. This can have real-world consequences. One such example of this is the "infodemic" surrounding coronavirus disease (COVID-19) vaccinations that spread rapidly on social media platforms, reducing vaccination rates in communities of color.

Various efforts are being made to combat the spread of misinformation and disinformation on digital platforms. For example, some have suggested using blockchain technology to track and verify sources, while others are promoting the labeling of websites according to their trustworthiness or incorporating digital literacy as part of the curriculum.

The Filipino journalist and Nobel Peace Prize laureate Maria Ressa launched a 10-point action plan to address the information crisis, calling on democratic governments to require tech companies to carry out and publicize independent human rights impact assessments, protect media freedom and safeguard journalists' safety, and approve legislation banning surveillance advertising.

The Swiss-based International Panel on the Information Environment (IPIE) is a newly formed global science organization, bringing together hundreds of leading scientists committed to providing actionable scientific knowledge about threats to the world's information environment. The mission of the IPIE is to provide policymakers, industry, and civil society with independent scientific assessment on the global information environment by organizing, evaluating, and elevating research, with the broad aim of improving the global information environment.

Sources: S. Lee Myers. 2023. With Climate Panel as a Beacon, Global Group Takes on Misinformation. *The New York Times*. 24 May; PricewaterhouseCoopers. Disinformation Attacks Have Arrived in the Corporate Sector. Are You Ready? ; The Nobel Prize. Nobel Prize Summit: Truth, Trust and Hope; and International Panel on the Information Environment. Scientists.

---

[288] C. Newton. 2023. Microsoft Lays off Team That Taught Employees How to Make AI Tools Responsibly. *The Verge*. 14 March. https://www.theverge.com/2023/3/13/23638823/microsoft-ethics-society-team-responsible-ai-layoffs.

[289] Although companies are increasingly making formal commitments to human rights, there is a wide gap between rhetoric and action, with 60% saying they support rights governance, 17% disclosing remedy procedures for rights harms, and only 10% conducting credible, regular due diligence through human rights impact assessments. Only Microsoft earns a perfect score for its transparency on assessing the impact of government policies on freedom of expression and privacy.

will appear, who can purchase ads, and how prices are set. They should also be required to report on their progress toward linguistic equity and conduct human rights impact assessments on their ad policies.

The future of accountability in the technology sector is closely tied to investors uniting around rights-based approaches to assess environmental, social, and governance (ESG)-related risks to society that tech companies are either generating or enabling. Among shareholders, the RDR standards are gaining broad acceptance for evaluating tech companies' performance. In 2022, investors representing $9 trillion in assets signed an Investor Statement on Corporate Accountability for Digital Rights,[290] challenging platform companies ranked by RDR to make specific improvements (also Box 9).

Growing advocacy in support of digital human rights is also coming from cities. With 70% of the world's population expected to live in urban areas by 2050, cities are expanding their capacity for digital transformation strategies that are people centered and support sustainable urbanization. The deployment of digital solutions in cities must be based on transparency, accountability, and community participation, especially in geographies where considerations of inclusion in technology and ethical concerns are still nascent.

## Box 9: What Is a Human Rights Impact Assessment in the ICT Sector?

Human rights impact assessments (HRIAs) are still a relatively novel concept for many companies operating in the information and communication technology (ICT) and digital media space for assessing human rights risks associated with their products, services, and actions. HRIAs can play a critical role to inform and guide corporate strategies, especially in challenging markets. While companies should publicly commit to undertaking HRIAs, experience has shown that many reports remain confidential, or, if disclosed, are partly redacted to protect sensitive information, and avoid jeopardizing operations against potential retaliation. An HRIA in the ICT and digital media space should cover at a minimum the following topics:

- The international legal and moral foundations for the rights to freedom of expression and privacy (Universal Declaration of Human Rights, United Nations Framework on Business and Human Rights).
- The general human rights landscape in the country or region, with a focus on the rule of law, freedom of expression, and privacy. Companies should consider the human rights implications just as in-depth as their due diligence on the local business climate and regulation.
- Local laws regarding free expression and privacy. Companies should have a clear sense of what the law, regulations, court decisions, and administrative practices in a local jurisdiction require of local businesses regarding the protection of free speech and privacy.
- Business plans for market entry. To map the company's products to potential risk areas, the team conducting the HRIA should be familiar with the business plans to be able to develop strategies to limit human rights risks.
- The potential to promote human rights. To the extent an ICT company's products have the potential to promote social good and human rights, the company should explore opportunities in the HRIA.
- Risk scenarios based on the company's products and operations. The team leading the HRIA should assess possible intersection points between the business and human rights issues.
- Proposed strategies for mitigating risks and protecting human rights. This will require technical detail about system architecture and jurisdictional choices. For instance, an ICT company may limit access by local staff to user data. If feasible, a company may locate data servers with sensitive information in a jurisdiction with stronger rule of law.

Source: Ranking Digital Rights. Human Rights Impact Assessment.

---

290  Investor Alliance for Human Rights. Investor Statement on Corporate Accountability for Digital Rights. https://investorsforhumanrights.org/investor-statement-corporate-accountability-digital-rights-0.

The Cities Coalition for Digital Rights, together with UN Habitat and United Cities and Local Governments, recently released a declaration[291] that addresses digital technologies, data, connectivity, and participatory governance and calls for:

- improved policies, laws, plans, and strategies for better open and ethical digital service standards;
- improved access to affordable and accessible internet and digital services on equal terms, as well as the digital skills to make use of this access and overcome the digital divide;
- improved privacy and control over personal information through data protection in both physical and virtual places;
- increased understanding of the technological, algorithmic, and AI systems that impact their lives, and the ability to question and change unfair, biased, or discriminatory systems; and
- transparent opportunities to shape the technologies designed for them, including managing our digital infrastructures and data as a common good.

The Digital Rights Governance Framework outlines how cities can uphold a human rights-based approach with regard to the digitalization of their services.[292] Its *foundation* should reflect a city's core values (e.g., respect for the right to privacy in online spaces, freedom of expression, emphasis on inclusivity and community, accessibility of data solutions) and be made explicit in its public stance on key thematic issues (e.g., fair treatment and equal access to prevent differentiation among people, residents' voice in how data collected about them are utilized, state protection against human rights abuses by third parties). These commitments to a digital bill of human rights and data policy with sovereignty at a local level are being adopted by major metropolitan governments (e.g., London, Barcelona, Los Angeles) and signal tangible actions (e.g., City of Portland prohibits the use of facial recognition technologies by private entities in public places).

To integrate these digital human rights commitments into the city's work, a whole-of-government approach is needed. This could be formalized by appointing elected officials, mainstreaming digital human rights in all policy areas, creating sandboxes and living labs to test new policy measures and technologies, or building up a repository of human rights impact assessments for new digital technology deployments.

The implementation of digital rights and services requires *tools*. City governments can take positive measures on digital inclusion, equality, and equity (e.g., digital gap surveys); transparency and accountability (e.g., assess risks, harm, and benefits of data use; conduct algorithmic auditing); community participation; privacy, safety, security, and protection (e.g., de-identify personal data); procurement (e.g., new models for technology acquisition); capacity building for local data modeling and use; and provision of public goods (e.g., open-government data). A four-city pilot of the framework was launched in 2022.[293]

A well-publicized case study involves the development of Toronto's Digital Infrastructure Plan. This initiative attracted global attention in anticipation of a significant private sector role in developing a smart city along Toronto's waterfront for 100,000 residents at a cost of $1.3 billion ("Sidewalk Toronto"). Residents, privacy

---

291  Cities Coalition for Digital Rights. Declaration. https://citiesfordigitalrights.org/declaration.
292  Cities Coalition for Digital Rights. Digital Rights Governance Framework. https://citiesfordigitalrights.org/framework.
293  Cities Coalition for Digital Rights. 2022. Brussels, Dublin, Sofia and Tirana selected to pilot a digital rights governance framework. Press release. 25 March. https://citiesfordigitalrights.org/selectedcities

advocates, and the city council flagged the lack of a city-level digital policy framework and insisted that privately collected data (e.g., on air quality, building and site-level systems, retail patterns, pedestrian, and vehicle movements) should remain in the public domain. While the private tech company withdrew from the project over these objections, it galvanized the adoption of the digital infrastructure plan.[294] The plan reflects the key principles of equity and inclusion; privacy and security; digital autonomy; democracy and transparency; and shared social, economic, and environmental benefits in a well-run city.

Another example involves the role of community networks in Latin America, where an estimated 250 million people lack internet connectivity, digital literacy, and access to digital public services. In this landscape, in which low-income households may not attract private investments and have not benefited from government subsidies, community networks have emerged as viable alternatives to connect to the online environment. These bottom-up networks are owned by locals, who jointly manage it as a common good for nonprofit purposes. They respond to demands for infrastructure, create new governance models for information and technology access, and promote better alignment with people's aspirations—a model that can probably be applied in any region.[295]

# 8.2. Digitally Vulnerable Groups

The digital divide remains a defining experience for many. According to the UN, almost half the world's population, i.e., 3.7 billion people, most of them women, and most of them living in developing countries, are still offline. The COVID-19 pandemic deepened concerns that the rapid expansion of digital services for education, health, social services, and job opportunities was not experienced equally. It revealed a deep digital gulf in the provision of meaningful connectivity and digital uptake and usage, leading to increasing inequalities between those who have access and can use these opportunities, and those without such possibilities, be it because they are too poor, marginalized, or live in rural areas without reliable electricity and internet connectivity.

The pandemic made it clear that digital transformations are interconnected with broader systemic shifts, including supply chain disruptions and geopolitical rivalry, and will pose new risks for society and a country's growth trajectory. It has also raised the specter of the "datafication" of personal information, algorithmic reinforcement of discrimination against structurally disadvantaged groups, and the influence of social media platforms in channeling misinformation and disinformation.

This section will explore how digital risks are affecting two vulnerable groups: school-age children as well as refugees and migrants.

---

294 A. Bozikovic. 2022. An Inside Look at How Alphabet's Sidewalk Labs Failed in Toronto. *Architectural Record*. 22 March. https://www.architecturalrecord.com/articles/15573-the-end-of-sidewalk-labs.

295 Association for Progressive Communications. 2021. Community Networks from Asia, Africa and Latin America Gather Online to Share Experiences, Learnings and Resources. News release. 5 August. https://www.apc.org/en/news/community-networks-asia-africa-and-latin-america-gather-online-share-experiences-learnings-and.

## 8.2.1 Children

The digital environment has become one of the most frequented spaces of daily life, blurring long-standing distinctions between school, work, and home. A key factor in the evolving digital risk landscape revolves around the increasing use of digital devices and technologies to enable education, knowledge acquisition, communication, social connection, and civic discourse. Digital service providers and platform companies, whose business models rely on maximizing user engagement, play a key role in this transformation.

The digital risk landscape for education systems, software applications, and data is particularly complex, multifaceted, and rapidly evolving. During the COVID-19 pandemic, most edtech products were offered at no direct financial cost, allowing governments to offload the actual costs of online education onto parents. Merging public education with for-profit learning platforms, video conferencing, and online proctoring services posed new risks to students' privacy, who were unknowingly forced to pay for their learning with their data, potentially compromising their rights to privacy, providing access to sensitive information, and potentially limiting their freedom of thought.[296] Current consent models of school-based digital platforms often present a false choice, as opting out may exclude students from learning opportunities and parents from important messages sent out by the school.

The widespread practice of *un-bundling* and *re-bundling* of education services can elevate the private good at the expense of the public good and lead to a monopolization of the education sector by a few companies.[297] Universities, edtech companies, and investors have used market-led approaches to monetize different aspects of the education experience, including student support, credentials, networks, curriculum, learning pathways, resources, and academic expertise. A cautionary example is the replacement of free massive open online courses with a "freemium" model, in which the lack of professional academic support for learners raises equity concerns and casts doubts over the promise to deliver education-for-all programs.

The increasing commercialization of education—which is in the interest of many technology providers— poses significant risk of missing out on adjusting educational content to the social and cultural context. Applications are hard to adapt without broad stakeholder participation and potentially with policy changes. As edtech corporations expand their sales globally, this can create asymmetries in power when local data is extracted from host countries potentially creating data protection and security challenges, or by drawing education systems into adopting a narrow set of products across all institutions and imposing their pedagogical approaches through digitalized tutoring systems.[298]

Students are facing commercial risks due to fraudulent or misleading claims. A 49-country review of 164 educational apps and websites in 2022 found that nearly 90% of the educational tools were designed to pass information on to advertising technology companies to target students' interests and influence buying preferences. Malware downloads and social networking apps open a backdoor to students' personal

296   H. J. Han 2022. How Dare They Peep into My Private Life? *Human Rights Watch*. 25 May. https://www.hrw.org/report/2022/05/25/how-dare-they-peep-my-private-life/childrens-rights-violations-governments.

297   L. Czerniewicz. 2028. Unbundling and Rebundling Higher Education in an Age of Inequality. *Educause Review*. 29 October. https://er.educause.edu/articles/2018/10/unbundling-and-rebundling-higher-education-in-an-age-of-inequality.

298   T. Adam. 2019. Digital Neocolonialism and Massive Open Online Courses: Colonial Pasts and Neoliberal Futures. *Learning, Media and Technology*. 44 (3): 365–80, https://doi.org/10.1080/17439884.2019.1640740.

information being used for profiling, affecting their fundamental legal rights and freedoms.[299] Children may not comprehend the motivation behind this type of data collection or the longer-term privacy consequences. Few OECD countries have adopted laws that specifically address consumer risks for school-age children.[300] The EU Digital Services Act which is becoming effective over the course of several years, addresses some of the risks to consumers. It does, for example, prohibit the use of personal data by online platforms to show targeted advertisements to children and teenagers.[301] The EU has levied significant fines on the mishandling of children's data and improper data transfers in breach of GDPR, including a $1.3 billion penalty for Meta in May 2023.

Parents, teachers, and school administrators need to learn more about how edtech companies are using student data. Violations of data privacy and protection arise not only from students' personal information they, their parents, or friends knowingly share ("data given"), but also involves information gleaned from their online activities via data tracking technologies ("data traces") as well as from data derived from algorithms ("inferred data"). Yet, the capacity of parents and teachers to effectively understand and supervise the digital activities of students is limited, with many lacking the critical digital literacy skills to protect children against these risks.

Young users face a wide range of content-related risks in the digital environment. They may be exposed to content that can be (i) hateful, motivated by the victim's religion, race, gender, disability, or sexual orientation; (ii) harmful, such as online scams, pornography, or violent material; (iii) illegal to distribute, subject to a country's laws and sociocultural norms; or, increasingly, (iv) meant to misrepresent news and information. With increased social media use contributing to poorer reading and shorter attention spans, students are more vulnerable to believing fake news and may lack the ability and knowledge to identify misleading information. Finland, which is ranked first in Europe on resilience against misinformation, is pursuing a concerted effort to teach students about fake news. Starting in preschool, media literacy is part of the national core curriculum.[302]

School-age children may also actively participate in or are exposed to conduct and contact risks through hateful or harmful encounters in the digital environment. Cyberbullying can lead to negative consequences for the victim's personal development and safety. The exchange of sexual messages and images ("sexting") is a growing practice among teens, with adverse implications for individual privacy, health, and well-being, and potential criminal consequences.[303] This risk category also includes sexual exploitation and cyber grooming via email, chat rooms, and social media platforms.

[299] UK Information Commissioner's Office. Age-appropriate design: a code of practice for online services. https://ico.org.uk/for-organisations/uk-gdpr-guidance-and-resources/childrens-information/childrens-code-guidance-and-resources/age-appropriate-design-a-code-of-practice-for-online-services/12-profiling/.

[300] OECD. 2020. Protecting children online: An overview of recent developments in legal frameworks and policies. *OECD Digital Economy Papers*. No. 295. Paris: Organisation for Economic Co-operation and Development. https://doi.org/10.1787/9e0e49a9-en.

[301] European Commission. The Digital Services Act: Ensuring a safe and accountable online environment. https://commission.europa.eu/strategy-and-policy/priorities-2019-2024/europe-fit-digital-age/digital-services-act-ensuring-safe-and-accountable-online-environment_en.

[302] J. Gross. 2023. How Finland Is Teaching a Generation to Spot Misinformation. *The New York Times*. 10 January. https://www.nytimes.com/2023/01/10/world/europe/finland-misinformation-classes.html.

[303] C. Doyle, E. Douglas, and G. O'Reilly. 2021. The Outcomes of Sexting for Children and Adolescents: A Systematic Review of the Literature. *Journal of Adolescence*. 92: 86–113. https://doi.org/10.1016/j.adolescence.2021.08.009.

Where affordable connectivity and digital devices have enabled school-age children easy access to digital media, this exposure has raised widespread public concerns for their health and well-being. There is growing demand to better understand the effects of digital media use on brain function and structure, as well as on physical and mental health, education, social interaction, and politics.[304] Intensive digital media use has been implicated in reducing working memory capacity; in psychological problems, from depression to anxiety and sleep disorders; and in influencing the level of text comprehension while reading on screens. Higher use levels are associated with lower happiness levels, especially for girls. Heavy digital media multitaskers among teenagers are found to have poorer memory function, increased impulsivity, less empathy, and a higher amount of anxiety.[305] Early extensive screen use in preschoolers can have dramatic influences on language networks. Cyberbullying can adversely affect the affected child's mental health, including higher levels of depression, anxiety, and social exclusion.[306]

While the purported benefits of AI in education suggest a future of virtual teachers and individualized learning, fundamental issues remain unaddressed. There is a risk that AI in education systems will be deployed without an ethical framework that uses learning and human development as a starting point and explicitly considers issues such as fairness, accountability, transparency, bias, autonomy, agency, and inclusion. New emotion recognition capabilities, for instance, not only can give an idea of "who you are" but also provide information about students' mental state ("how you feel") without their awareness and consent, prompting concerns of misclassification and stigmatization. A second concern is that most AI in education tools tend to drive the homogenization of students. The risk is that the ability of teachers to personalize their teaching in response to each student may be replaced by "individual" pathways that are based mostly on averages of prior learners, while collaboration and other social interaction aspects of teaching and learning will be ignored, thus replicating existing biases. The OECD recommends that unless appropriate measures are in place to protect children from any harmful effects, digital platforms and applications should not allow the profiling of children.

The opportunities and risks that children and adolescents face in the digital environment cut across borders and jurisdictions and require global and regional governance coalitions. Clear guidelines and firm policy and compliance frameworks are needed to enable equitable access for all students. The OECD recently issued recommendations to find a balance between protecting children from risk and promoting the opportunities and benefits that the digital environment can provide. Several guidance documents have been released to protect students' data and fundamental rights in educational settings, including by the Council of Europe, the International Telecommunication Union (ITU), and the Global Privacy Assembly.[307]

[304] M. Korte. 2020. The Impact of the Digital Revolution on Human Brain and Behavior: Where Do We Stand? *Dialogues in Clinical Neuroscience*. 22 (2): 101–11. https://doi.org/10.31887/DCNS.2020.22.2/mkorte.

[305] E. Barry. 2023. Social Media Use Is Linked to Brain Changes in Teens, Research Finds. *The New York Times*. 3 January. https://www.nytimes.com/2023/01/03/health/social-media-brain-adolescents.html.

[306] C. Zhu et al. 2021. Cyberbullying Among Adolescents and Children: A Comprehensive Review of the Global Situation, Risk Factors, and Preventive Measures. *Frontiers in Public Health*. 9. https://www.frontiersin.org/articles/10.3389/fpubh.2021.634909.

[307] Council of Europe. Children's Data Protection in an Education Setting-Guidelines 2021. https://edoc.coe.int/en/children-and-the-internet/9620-childrens-data-protection-in-an-education-setting-guidelines.html.; International Centre for Missing & Exploited Children. ITU Child Online Protection Guidelines 2020. https://www.icmec.org/itu-child-online-protection-guidelines-2020/.; and Global Privacy Assembly. Adopted Resolutions. https://globalprivacyassembly.org/document-archive/adopted-resolutions/.

Child rights impact assessments have been adopted in Australia; Canada; Europe (Belgium, Bulgaria, Finland, Scotland, Spain, Sweden); New Zealand; and more recently, in partnership with UNICEF in developing countries (including in Bolivia, Colombia, Costa Rica, El Salvador, India, Malaysia, Rwanda, South Africa, and Tanzania). The Council of Europe's recent recommendations on the rights of the child in the digital environment calls on member states to "require business enterprises to perform regular child-rights risk assessments for digital technologies, products, services, and policies and to demonstrate that they are taking reasonable and proportionate measures to manage and mitigate such risks."[308] To date, few global internet and mobile companies' practices meet minimum standards in the areas of privacy, freedom of expression, and remedy mechanisms for reported harm. There are exceptions: LEGO pledged compliance with the Children's Rights and Business Principles.[309]

## 8.2.2. Refugees and Migrants

Digital identity and biometrics[310] have long been divisive topics in the humanitarian sector.[311] On the one hand, governments and development organizations around the world have deployed foundational digital identities, often with biometric verification, to register citizens and non-citizens alike. The issuance of functional digital IDs can increase inclusion and ensure access to social protection, education, health care, or financial services.

On the other hand, the use of biometrics has raised serious human rights concerns around choice, questions about inclusion and exclusion, informed consent, as well as data privacy and protection for those in need of humanitarian assistance. Increasingly, exclusion from digital ID systems means exclusion from aid, as biometrics are often tied to access to services in displacement settings, particularly for cash transfers. By and large, expectations around the efficiency of biometrics have not been met, particularly for end users, raising doubts whether the benefits to the providers of aid or social protection services are worth the trade-off of a more difficult experience for affected communities.

---

[308] Council of Europe. Guidelines to respect, protect and fulfil the rights of the child in the digital environment. https://rm.coe.int/guidelines-to-respect-protect-and-fulfil-the-rights-of-the-child-in-th/16808d881a.

[309] LEGO Group. *Responsibility Report 2018*. www.lego.com/cdn/cs/aboutus/assets/blt9f56973f588882fd/Responsibility-Report-2018.pdf.

[310] Biometrics include physiological characteristics (e.g., fingerprints, facial structures, iris patterns, voice recognition) that are measured and assessed for either identification or verification.

[311] K. Holloway, R. Al Masri, and A. Abu Yahia. 2021. Digital Identity, Biometrics and Inclusion in Humanitarian Responses to Refugee Crises. *ODI*. 6 October. https://odi.org/en/publications/digital-identity-biometrics-and-inclusion-in-humanitarian-responses-to-refugee-crises/.

Due to a lack of awareness and/or capacity, there have also been cases of improper handling or exfiltration of sensitive information. Such was the case of the World Food Programme which conducted privacy impact assessments only after or that of the International Committee of the Red Cross (ICRC) which suffered a cyber attack by an unknown state actor that compromised the personal data of 515,000 vulnerable people on a third-party server.[312]

Another source of vulnerability concerns the expanding array of digital technologies being deployed as part of surveillance and data-driven refugee and immigration policies.[313] Every interaction migrants have with immigration authorities and enforcement agencies entails the collection and processing of personal data. Yet, there are limited safeguards to regulate and oversee the use of technologies and processing of data. Governments are increasingly using migrants' electronic devices to corroborate information by using mobile extraction tools, which are able to download key data. AI tools are used in immigration settings to profile people based on population-scale data and automate decision-making (e.g., visa applications, identification of refugees). The use of facial recognition technologies for so-called "frictionless surveillance" is beset by misidentification errors, especially among minority groups. The transfer of border controls and technologies to foreign countries, specialized agencies, and private security companies ("externalization of borders") has become a preferred instrument to stop migratory flows, which opens the door to human rights violations and corruption, and can undermine democratic governance.

As the use of digital identity and biometrics continues to grow, governments and the humanitarian and the development sectors must find new ways to improve the systems that are in place and mitigate potential risks. Accountability to affected people should include responsible data collection and data protection policies. Where biometrics are used, inclusivity and equal access for all should be ensured through human-centered design methods. Users should have a choice in whether their biometrics are captured and used ("informed consent"). Their desire for privacy and ethical respect should be consistent with how much data are collected, how they are protected, and with whom they are shared. It is not obvious that end users understand the technology being used and the extent to which their data will be used, protected, and shared; nor can they opt in or out without affecting the level of aid or services they receive. For organizations using biometrics, the primary aim should be to minimize the amount of data being collected and delete them once it has served its purpose.

[312] International Committee of the Red Cross. 2022. Cyber-Attack on ICRC: What We Know. 21 January. https://www.icrc.org/en/document/cyber-attack-icrc-what-we-know.

[313] Privacy International. 10 Threats to Migrants and Refugees. http://privacyinternational.org/long-read/4000/10-threats-migrants-and-refugees.

**A balancing act.** Navigating the risks of digital technology in education to ensure a safe and effective learning environment (photo by ADB).

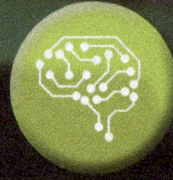

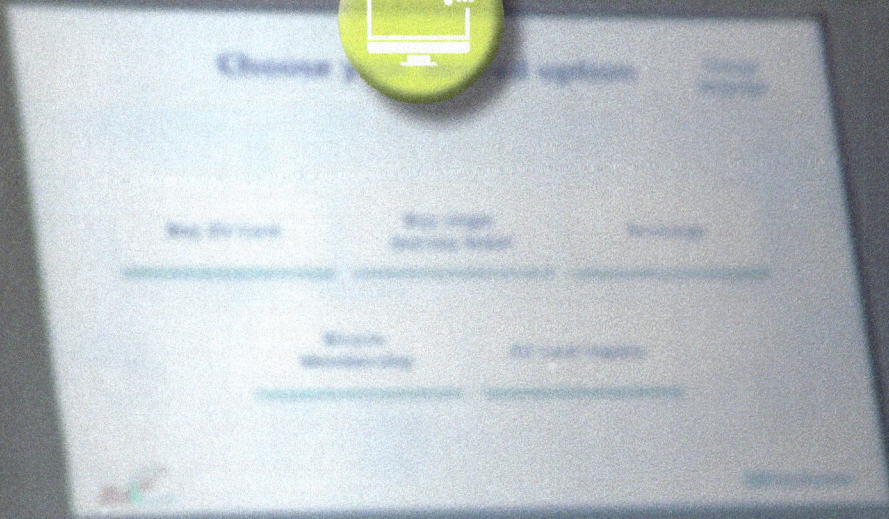

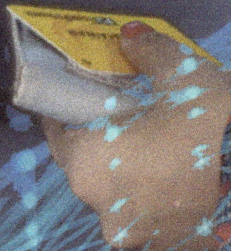

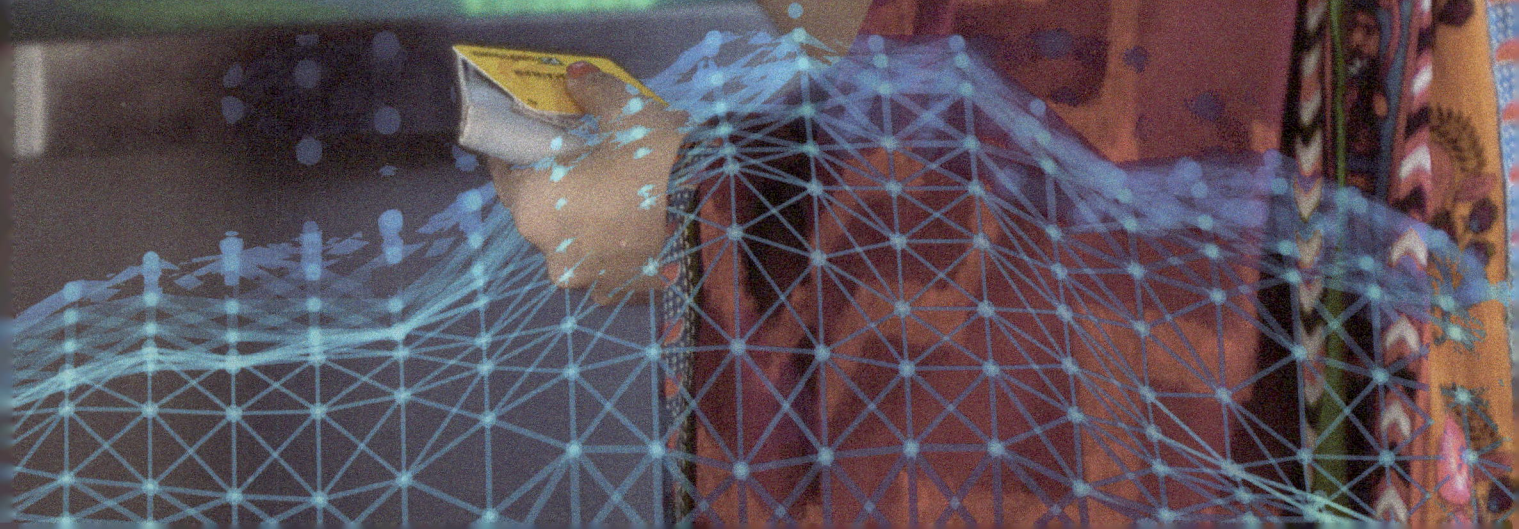

# 9

## SUSTAINABILITY RISKS

Digital technologies have the potential to contribute to decarbonization across sectors (e.g., energy, mobility, industry) and to promote circular and shared economies, dematerialization, resource and energy efficiency, monitoring and conservation of ecosystems, protection of the global commons, and sustainable behaviors.[314] However, this positive impact is not a given.

In fact, digitalization itself is responsible for a large environmental footprint. Since 2010, the doubling of internet users globally has fueled a 12-fold increase in internet traffic; estimates are that 70% of new economic value created over the next decade will be based on digital platforms.[315] If the internet was a country, it would be the sixth-biggest electricity consumer on the planet, consuming 7% of global electricity[316] and generating up to 3.7% of global greenhouse gas emissions.[317]

## 9.1. Linking Digital Services and Energy Demands

Although digital technologies are often overlooked as primary carbon producers, their impact on global sustainability is widespread, as are their points of origin. There is growing interest in understanding how the environmental footprint of digital infrastructure, devices, and user applications, model the threats to the planet and future generations, and develop sustainable solutions toward global digitalization.

A systems framework helps visualize the linkages and contributing factors, both positive and negative, arising from the growing demand for digital services, which have led many governments, industries, and consumer groups to target specific areas as part of their commitment to achieving net-zero emission targets by 2030 or even earlier. The basic framework and choices are summarized in Figure 28.[318] Demand for digital services drives production of digital devices and increases energy use, leading to an increase in overall energy demand and an increase of end-of-life electronic waste (e-waste). As long as the electricity supply is based on fossil fuels, consumption and use increases greenhouse gas emissions, worsening global warming and climate change that, in turn, are adversely impacting human health. Even renewable energy sources have an emissions footprint and environmental and social impact that needs to be considered.

This note focuses on three specific areas of the digital ecosystem that have a disproportionately large impact on energy consumption, pollution, and human well-being: the hardware supply chain, data centers, and cryptocurrency mining.

---

[314] International Institute for Applied Systems Analysis. The Digital Revolution and Sustainable Development: Opportunities and Challenges: Report Prepared by The World in 2050 Initiative. https://doi.org/10.22022/TNT/05-2019.15913.

[315] World Economic Forum. Shaping the Future of Digital Economy and New Value Creation. https://www.weforum.org/platforms/shaping-the-future-of-digital-economy-and-new-value-creation/.

[316] F. Bordage. 2019. The Environmental Footprint of the Digital World.

[317] Climate Care. The Carbon Footprint of the Internet. https://www.climatecare.org/resources/news/infographic-carbon-footprint-internet/.

[318] United Nations Environment Programme. The Growing Footprint of Digitalisation. http://www.unep.org/resources/emerging-issues/growing-footprint-digitalisation.

**Figure 28: A Systems Perspective for the Linkages between Digital Services and Energy Demands**

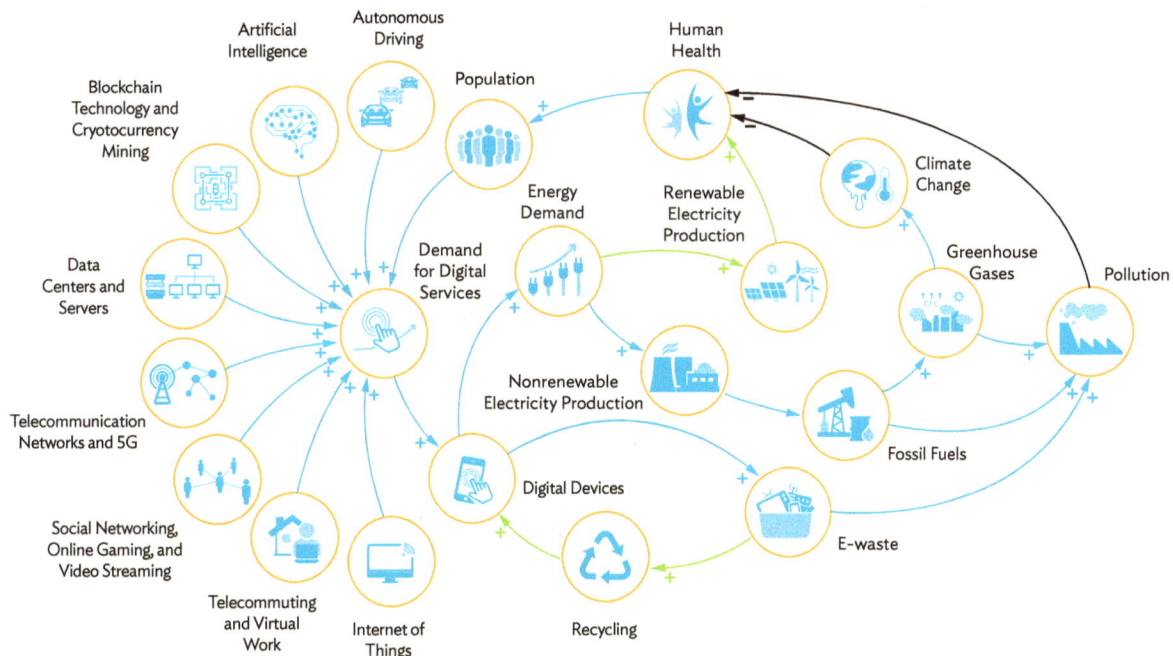

Source: United Nations Environment Programme. The Growing Footprint of Digitalisation.

## 9.2. The Supply Chain Life Cycle for Digital Products and Energy Technologies

Starting upstream, the extraction of metals and rare earth minerals used to produce cell phones, computers, and servers generates its own ESG footprint. Many of the global reserves of the key 18 scarce metals and rare earth elements, including cobalt, graphite, and copper, tend to be concentrated in countries struggling with political instability, where weak governance in the mining sector can be linked to violence, corruption, human rights abuses, and severe environmental damage.

On the downstream side, the lifetime of digital hardware is often only a few years, driven by rapid technological advancement and business models that promote frequent equipment upgrades. The challenges of proper collection, recycling, and reuse are immense. In 2019, a record 53.6 million metric tons of e-waste was produced—the equivalent weight of 125,000 Boeing 747 jumbo jets. This makes e-waste the world's fastest-growing domestic waste stream, fueled mainly by higher consumption rates of electric and electronic devices, short life cycles, and few options for repair. Less than 20% of e-waste is formally reported as collected and recycled. The remaining 80% either ends up in landfills or is being informally recycled—much of it by hand in developing countries, exposing workers to hazardous and carcinogenic substances, and contaminating soils and groundwater.[319]

---

[319] Geneva Environment Network. The Growing Environmental Risks of E-Waste. https://www.genevaenvironmentnetwork.org/resources/updates/the-growing-environmental-risks-of-e-waste/.

According to global estimates for 2019, the value of e-waste raw materials is around $ 57 billion, making it an important source for the mining of secondary raw materials. The Asia and Pacific region has begun to recognize the importance of proper e-waste management. Cambodia, India,[320] and the PRC are among the 78 countries that have adopted e-waste legislation, with several other countries considering similar legislation. The e-waste management systems found in the Asia and Pacific region range from very advanced, such as in the Republic of Korea, Japan, and the PRC, to informal sector activities for collection, dismantling, and recycling.[321]

Given growing demand for digital devices and services, what policy actions are being considered? First, there are growing pressures on governments, mining companies, and financing institutions to improve the governance of ICT supply chains, especially regarding the extraction of rare earth minerals and metals, the recycling of e-waste, and the safe disposal of toxic material.[322] Second, given the limited ability to expand rare earths mining, novel methods for the recovery and separation of rare earth elements are maturing.[323] Third, evolutionary design and circular economy models could extend the life span of devices and enable the upgrading and replaceability of key components.

A 2019 report by the World Economic Forum[324] calls for a new vision for e-waste based on the circular economy concept, whereby a regenerative system can minimize waste and energy leakage.[325] The report supports the work of the E-waste Coalition, consisting of the International Labour Organization, ITU, United Nations Environment Programme, United Nations Industrial Development Organization, United Nations Institute for Training an Research , United Nations University, and secretariats of the Basel and Stockholm Conventions, which gather evidence at global and regional levels, case studies, and scenarios.[326] The EU is currently developing the concept of a digital "product passport," which will provide digital information on a product's origin, durability, composition, environmental and carbon footprint, reuse, repair and dismantling possibility, and end-of-life handling.[327] The Global System for Mobile Communications Association's ClimateTech program has created a legislative framework map of e-waste policies for 76 countries and published an Eco Rating to encourage suppliers and consumers to reduce the environmental impact of their mobile phones.[328]

[320] S. Arya and S. Kumar. 2020. E-Waste in India at a Glance: Current Trends, Regulations, Challenges and Management Strategies. *Journal of Cleaner Production.* 271: 122707. https://doi.org/10.1016/j.jclepro.2020.122707.

[321] V. Forti et al. 2020. The Global E-Waste Monitor 2020.

[322] CEE Bankwatch Network. 2021. Raw Deal. https://bankwatch.org/publication/raw-deal.

[323] Z. Dong et al. 2021. Bridging Hydrometallurgy and Biochemistry: A Protein-Based Process for Recovery and Separation of Rare Earth Elements. *ACS Central Science.* 7 (11): 1798–1808, https://doi.org/10.1021/acscentsci.1c00724; and Sustainability Times. 2021. There's a Safer, Cleaner Way to Recover Rare-Earth Metals from Old Phones and Laptops. https://www.sustainability-times.com/sustainable-business/theres-a-safer-cleaner-way-to-recover-rare-earth-metals-from-old-phones-and-laptops/.

[324] World Economic Forum. A New Circular Vision for Electronics–Time for a Global Reboot. https://www.weforum.org/reports/a-new-circular-vision-for-electronics-time-for-a-global-reboot.

[325] World Economic Forum. Shaping the Future of Digital Economy and New Value Creation.

[326] United Nations Institute for Training and Research. Global E-Waste Monitors. https://ewastemonitor.info/.

[327] S. Guth-Orlowski. 2021. The Digital Product Passport and Its Technical Implementation. *Medium* (blog). 2 December. https://medium.com/@susi.guth/the-digital-product-passport-and-its-technical-implementation-efdd09a4ed75.

[328] GSMA Mobile for Development. ClimateTech. https://www.gsma.com/mobilefordevelopment/climatetech/.

Apple offers a case study for the "greening" of the electronics industry. Following its 2017 announcement to one day make products without taking from the Earth, Apple expanded its Zero Waste-to-Landfill program to its supply chain, ensuring all the waste created at final assembly sites was reused, recycled, composted, or, when necessary, converted into energy. This effort included the first supplier sites to pursue Zero Waste certifications for Apple's footprint in the PRC. Apple works with recycling partners to recover the maximum amount of materials using innovative recycling technologies like developing a line of robots that are specially designed to efficiently disassemble iPhone devices. Based on supplier reports, 10% of the total product Apple shipped in 2019 came from either recycled or renewable sources, one-third of which has been confirmed through third-party certifications.[329]

To estimate the generation of e-waste and assess its environmental impacts, two tools are frequently used:

- **Material flow analysis** is an effective decision support tool that can trace the complex route of materials flowing into recycling sites, understand issues and gaps in the value chain, and recommend appropriate management strategies to reduce environmental impact. Four countries, Ghana, India, the PRC, and Viet Nam, have been intensely studied on the environmental and health impacts of e-waste.[330]
- **Life-cycle assessment** is an effective and popular environmental management tool that can evaluate environmental impacts of a product, service, or system throughout its life cycle, i.e., starting from raw material acquisition to waste management, and can identify hotspots and potential for improvement. Life-cycle analysis quantifies all the energy and material inputs used and all the emissions and waste materials produced throughout the product's life cycle.

The Global System for Mobile Communications Association has developed a sustainability assessment framework to better understand the landscape of mobile network operator efforts in social and environmental sustainability across the industry, including on climate impact and risk, waste and e-waste, and responsible sourcing.[331] Deutsche Telekom, for instance, published a detailed climate strategy, committing to net zero for scope 1 and 2 emissions by 2025.[332] Malaysia's Axiata Group has created a net-zero carbon road map to reduce operational carbon emissions by 45% by 2030 from a 2020 baseline.[333] Singapore's Singtel has adopted measures to adapt to and build resilience to the longer-term risks of climate change. Its board risk committee reviews climate risks, while members of the management committee provide oversight and stewardship on Singtel Group's carbon reduction strategy.[334]

329  Fraunhofer Institute for Reliability and Microintegration IZM, Berlin, "Proceedings_EGG2020_v2.Pdf." accessed 11 February 2022. https://online.electronicsgoesgreen.org/wp-content/uploads/2020/10/Proceedings_EGG2020_v2.pdf.

330  S. V. Withanage and K. Habib. 2021. Life Cycle Assessment and Material Flow Analysis: Two Under-Utilized Tools for Informing E-Waste Management. *Sustainability*. 13 (14): 7939. https://doi.org/10.3390/su13147939.

331  GSMA. The GSMA Sustainability Assessment Framework 2021. https://www.gsma.com/betterfuture/resources/the-gsma-sustainability-assessment-framework.

332  Deutsche Telekom. 2020 Corporate Responsibility Report: Climate Strategy. https://www.cr-report.telekom.com/site20/management-facts/environment/climate-strategy.

333  Axiata. Axiata Sustainability – Advancing to Zero. https://sustainability.axiata.com/.

334  Singtel. Climate Change and Environment. https://www.singtel.com/about-us/sustainability/sustainability-at-singtel/environment.

## 9.3. Estimating the Global Energy Use of Data Centers

Data centers represent the backbone of an increasingly digitalized world.[335] Demand for their services is rising rapidly; data-intensive technologies such as AI, smart and connected energy systems, distributed manufacturing, and autonomous vehicles promise to increase demand further. Given that data centers are energy-intensive enterprises, these trends have implications for global energy demand.

In 2018, global data center energy use rose to around 205 terawatt-hours, equivalent to 1% of global electricity consumption. While this represents a 6% increase compared with 2010, the energy intensity of global data centers (expressed as energy use per compute instance) decreased by 20% annually over the same period. These improvements in energy intensity can be attributed to increased server efficiencies, greater server virtualization, and decreased energy use for data center infrastructure (i.e., cooling and power provisioning), which reflect an ongoing shift away from small-scale data centers toward larger and more energy-efficient data centers, including those built for large cloud computing hyper-scalers which often require cross-border data flows but are able to look for a location with ample and more sustainable power supply.

Should these trends continue to play out over the next few years, estimates indicate that there is sufficient energy efficiency potential to absorb the next doubling of data center compute instances in parallel with a negligible increase in global data center energy use. These findings stand in contrast to recent predictions of rapid and unavoidable near-term energy demand growth, some of which are based on specific locations (e.g., Ireland) as opposed to a global bottom-up modeling framework.

What are possible policy actions? First, encourage the promotion of energy efficiency standards (such as Energy Star certification for server hardware, storage, and network devices) and a shift to cloud services[336] where economically and institutionally feasible, for example by introducing requirements into public IT procurement programs. With green procurement gaining momentum, decisions on cloud versus on-premises data centers will increasingly favor cloud computing unless local conditions and use cases require on-premise infrastructure. Second, investments in new technologies (such as quantum computing, materials for ultrahigh density storage, increased chip specialization, liquid and immersion cooling technologies, and local reuse of heat generated by data center servers) are necessary to accelerate the pace of renewable energy adoption and reduce carbon intensity. And third, recognizing that greater public data and modeling capacities will be crucial to net-zero transition, policymakers should enact robust data collection and disclosure requirements, develop data center energy models, and disseminate best practices for analytics. A 2022 panel discussion on regulatory initiatives regarding data centers organized by Digital Europe provided an update on renewable energy, global energy use estimates, technical innovations, and ongoing efforts by data center firms.[337]

---

[335] E. Masanet et al. 2020. Recalibrating Global Data Center Energy-Use Estimates. *Science*. 367 (6481): 984–86. https://doi.org/10.1126/science.aba3758.

[336] Amazon Web Services. 2021. Why Moving to the Cloud Should Be Part of Your Sustainability Strategy. *AWS Public Sector Blog*. 25 October. https://aws.amazon.com/blogs/publicsector/why-moving-cloud-part-of-sustainability-strategy/.

[337] Digital Europe. 2022. Data Centres: Powering the Green and Digital Transition in Europe. YouTube webinar. 1:26.56. 12 February. https://www.youtube.com/watch?v=O4l25h1W9sQ.

# 9.4. CryptoCurrency Mining

A defining characteristic of cryptocurrency mining is that it is essentially a process in which new virtual money is created and granted to the "miner" of this virtual money. Cryptocurrencies like Bitcoin or Ethereum rely on blockchain technology to determine the validity of transactions without the need for a central monetary authority to monitor its system. In the proof-of-work (PoW) consensus mechanism,[338] each miner serves as an auditor for the system by competing to validate the transaction through energy- and computer-intensive operations and adding it to the distributed ledger of verified transactions. This results in vast amounts of computing power being dedicated to PoW.

According to the Cambridge Center for Alternative Finance, if Bitcoin were a country, its greenhouse gas emissions of an estimated 71.8 million tons carbon dioxide equivalent per year would be comparable to that of Austria, Cambodia, Kenya, or New Zealand.[339]

While the energy consumption is relatively easy to estimate for PoW cryptocurrencies, their carbon emissions are much harder to ascertain and increasingly subject to regulation. Based on verified and estimated mining locations, the Cambridge Center for Alternative Finance calculated that crypto mining used ~39% of renewable energy in September 2020, far lower than earlier estimates of ~70% which may be related to a frequent movement of mining locations between countries. Mining has become a highly lucrative business with the amount of computing power dedicated to the process running at record levels.

Since crypto can be mined anywhere, mining activities are concentrated in regions with low energy costs—often related to excess hydro capacity (such as in the south west region of the PRC during the rainy season, Iceland, Paraguay, or Scandinavia) or natural gas flaring released during oil extraction (such as in South Texas, North Dakota, or Siberia).[340] Following the PRC's ban on crypto mining in May 2021 and faced with mounting criticism, mining activity has scattered across the world in search of renewable energy sources to power their computers. Crypto mining has become an issue for crypto hubs in Sweden, Norway, Kazakhstan, and Texas, where authorities are increasingly concerned that a significant amount of renewable energy for crypto currencies is undermining efforts of moving away from fossil energy sources.[341] In addition, observers estimate that every year and a half or so, the computational power of mining hardware doubles, making older machines obsolete and generating more e-waste than many midsize countries.[342]

---

[338]  The PoW concept is used by Bitcoin, Dogecoin, Litecoin, and other coins. It was initially used by the second-largest cryptocurrency, Ethereum, which has begun using proof-of-stake instead in 2022, but still has previous versions circulating.

[339]  University of Cambridge Centre for Alternative Finance. Cambridge Bitcoin Electricity Consumption Index. https://ccaf.io/cbeci/index/comparisons. Note: Country comparisons are, for better or for worse, the most common type of comparison. They are frequently used in public debates but should be treated with caution.

[340]  N. Carter. 2021. How Much Energy Does Bitcoin Actually Consume? *Harvard Business Review*. 5 May. https://hbr.org/2021/05/how-much-energy-does-bitcoin-actually-consume.

[341]  E. Szalay. 2022. EU Should Ban Energy-Intensive Mode of Crypto Mining, Regulator Says. *Financial Times*. 19 January. https://www.ft.com/content/8a29b412-348d-4f73-8af4-1f38e69f28cf.; and D. Brown. 2021. Bitcoin Miners Break New Ground in Texas, a State Hailed as the New Cryptocurrency Capital. The Washington Post. 8 July. https://www.washingtonpost.com/technology/2021/07/08/bitcoin-mining-texas-electricity/.

[342]  J. Huang, C. O'Neill, and H. Tabuchi,. 2021. Bitcoin Uses More Electricity Than Many Countries. How Is That Possible? *The New York Times*. 3 September. https://www.nytimes.com/interactive/2021/09/03/climate/bitcoin-carbon-footprint-electricity.html.

**Figure 29:** Cambridge Bitcoin Electricity Consumption Index

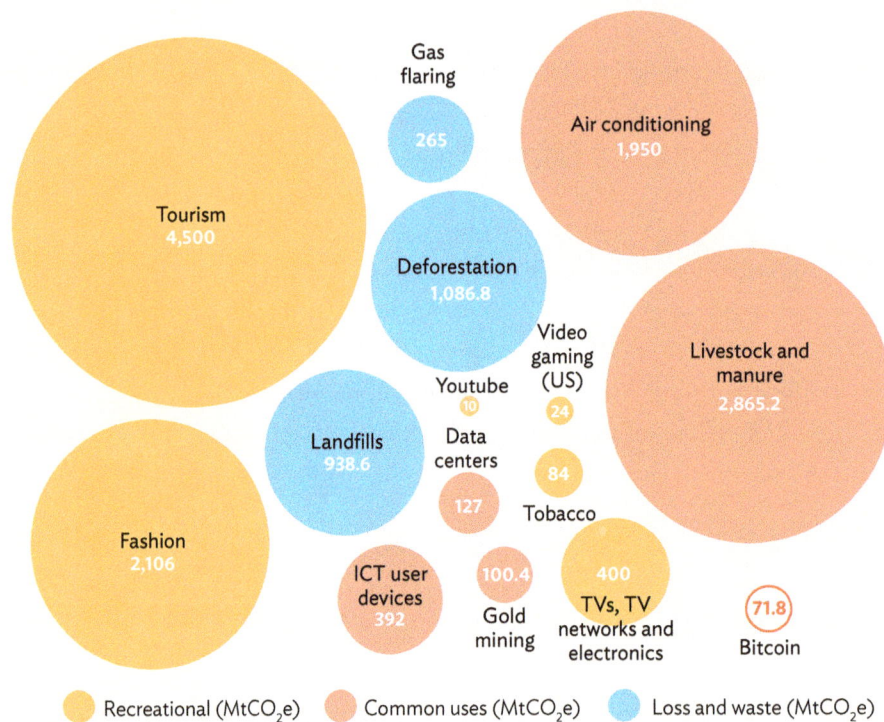

Gas flaring — 265

Air conditioning — 1,950

Tourism — 4,500

Deforestation — 1,086.8

Video gaming (US) — 24

Livestock and manure — 2,865.2

Youtube — 10

Landfills — 938.6

Data centers — 127

Tobacco — 84

Fashion — 2,106

ICT user devices — 392

Gold mining — 100.4

TVs, TV networks and electronics — 400

Bitcoin — 71.8

● Recreational (MtCO₂e)  ● Common uses (MtCO₂e)  ● Loss and waste (MtCO₂e)

ICT = information and communication technology, MtCO₂e = metric tons of carbon dioxide equivalent, US = United States.
Source: University of Cambridge Centre for Alternative Finance. Cambridge Bitcoin Electricity Consumption Index (extract as of 27 August 2023).

European Union regulators had considered provisions banning PoW cryptocurrencies as part of its deliberations of the draft Markets in Crypto-Assets regulation[343] in an attempt to push the industry toward the less energy-intensive "proof-of- stake" method. The proposal was eventually voted down,[344] but lawmakers called on the EU Commission to consider including PoW mining activities in the EU sustainable finance taxonomy.[345] Ethereum, the second-largest digital asset by market capitalization after Bitcoin, already migrated its main coin ETH to the proof-of-stake model in September 2022.[346]

On the private sector side, mining operators have launched initiatives like the Crypto Climate Accord—inspired by the Paris Climate Agreement—to commit to reducing Bitcoin's carbon footprint, adding incentives for miners to build out renewable energy technologies.[347] The largest push for decarbonization may, however, ultimately come from the side of institutional investors who are increasingly bound by stringent ESG rules and requirements.

343  European Union, Official Journal, "Regulation (EU) 2023/1114." https://eur-lex.europa.eu/legal-content/EN/TXT/PDF/?uri=CELEX:32023R1114 (accessed 27 August 2023).
344  M. Slater-Robins. 2022. The EU had a chance to ban Bitcoin – but chose not to. *TechRadar*. 15 March. https://www.techradar.com/news/the-eu-had-a-chance-to-ban-bitcoin-but-chose-not-to.
345  S. Handagama. 2022. Proposal Limiting Proof-of-Work Is Rejected in EU Parliament Committee Vote. *Coindesk*. 14 March. https://www.coindesk.com/policy/2022/03/14/proposal-limiting-proof-of-work-is-rejected-in-eu-parliament-committee-vote-sources/ (accessed 27 August 2023).
346  Ethereum. Proof-of-Stake (PoS). https://ethereum.org.
347  CryptoClimate. Crypto Climate Accord. https://cryptoclimate.org/accord/.

**Digital challenges.**
Security concerns in
cryptocurrency mining
(photo by ADB).

# 10

## DIGITAL
## RESILIENCE

## 10.1. Digital Resilience

A key take away from the COVID-19 pandemic is that the disruption and related trends will continue to affect people's lives, which is why building resilience has become more crucial than ever. In analyzing the global response to this crisis, another obvious lesson centers on the importance of digitalization and digital resilience as the world becomes more dependent on digital systems and technologies to cope, recover, adapt, and transform to this new reality.

The American Psychological Association frames resilience as both the process and the outcome of successfully adapting to difficult or challenging life experiences, especially through mental, emotional, and behavioral flexibility and adjustment to external and internal demands.[348] Resilience can be defined as the ability of nations, communities, organizations, and individuals to withstand, recover from, adapt to, and potentially transform amid change and uncertainty.[349] From this perspective, resilience should be seen as being more than just a defensive capability and include the ability to "bounce back better" or "bounce forward" by improving the physical, economic, social, political, and environmental conditions during the recovery process to further strengthen resilience and agility.

Digital resilience is often defined as the ability of IT systems to withstand and recover from threats such as cyber attacks and disasters. Noting the above, digital resilience can also mean using digital systems and technologies to strengthen people's resilience and agility in a local and society-wide context.

This chapter proposes a fundamental mindset shift away from a narrow view of risk management toward a holistic, people-centered approach to digital resilience. Without such a shift, future crises could further deepen poverty, hunger, and inequalities and could overpower humanity's recovery capacity. This broader perspective requires engagement with the entire digital ecosystem—the stakeholders, systems, and enabling environments that together make it possible for people and communities to use digital technologies to pursue economic and social opportunities—while taking steps to reduce risks.

## 10.2. The Accelerating Nature of Disruptions

The world is experiencing increasingly frequent, unpredictable, and unprecedented changes. The current era is characterized by the interplay of complex disruptions—climate change, health threats, geopolitical confrontations, food insecurity—with their disparate origins and long-term consequences (Figure 30).

Figure 31 profiles different types of shocks based on their impact, lead time, and frequency. The impact of a shock can be influenced by how long it lasts, its frequency, the cascading effects it has across geographies and industries, and whether a shock affects only the supply side or also hits demand.

---

348   American Psychological Association. Resilience. https://www.apa.org/topics/resilience.
349   R. Heeks. 2016. RABIT: A New Toolkit for Measuring Resilience. *Nexus for ICTs, Climate Change and Development*. 14 November. http://www.niccd.org/rabit-a-new-toolkit-for-measuring-resilience/.

**Figure 30: Charting the Increasing Frequency and Severity of Disruptions**

| IMF World Uncertainty Index[1] (thousand) | Federal Reserve Board Geopolitical Risk Index[2] | Companies Subject to a Cyber Breach per Year (%) | Disasters per Year (number) |
|---|---|---|---|

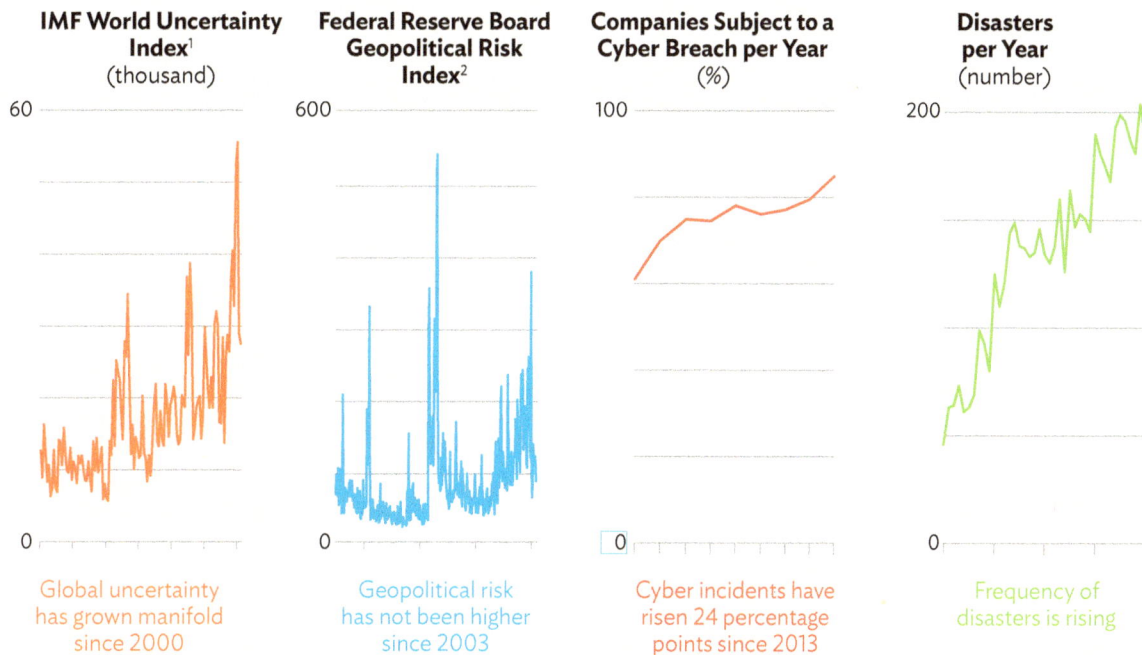

Global uncertainty has grown manifold since 2000

Geopolitical risk has not been higher since 2003

Cyber incidents have risen 24 percentage points since 2013

Frequency of disasters is rising

IMF = International Monetary Fund.
Notes:
[1] Based on the percentage of the word "uncertain" (or its variants) in the Economist Intelligence Unit country reports.
[2] Automated text-search results from the electronic archives of 11 newspapers. Index was calculated by counting the number of articles related to geopolitical risk in each newspaper for each month (as share of the total number of news articles).
Source: F. Nauck et al. 2021. The Resilience Imperative: Succeeding in Uncertain Times. *McKinsey*. 17 May.

Shocks can be categorized as *catastrophes* that cause trillions of dollars in losses, and as *disruptions* that are serious and costly but on a smaller scale. Some of these events are foreseeable and offer some degree of advance warning (e.g., financial crises, major military conflicts, pandemics), while others are unanticipated, even though they may occur with some frequency (e.g., hurricanes, common cyber attacks) or have not yet taken place (e.g., a cyber attack on foundational global systems).

Organizations tend to focus much of their attention on preparing for and managing the types of shocks they encounter most often ("anticipated disruptions"). Shocks that occur with increasing regularity such as supply disruptions or common cyber attacks are, to some extent, preventable. By contrast, organizations find it more challenging to prevent external disruptions, such as pandemics and disasters, other than to analyze their exposure and put different resilience measures in place.

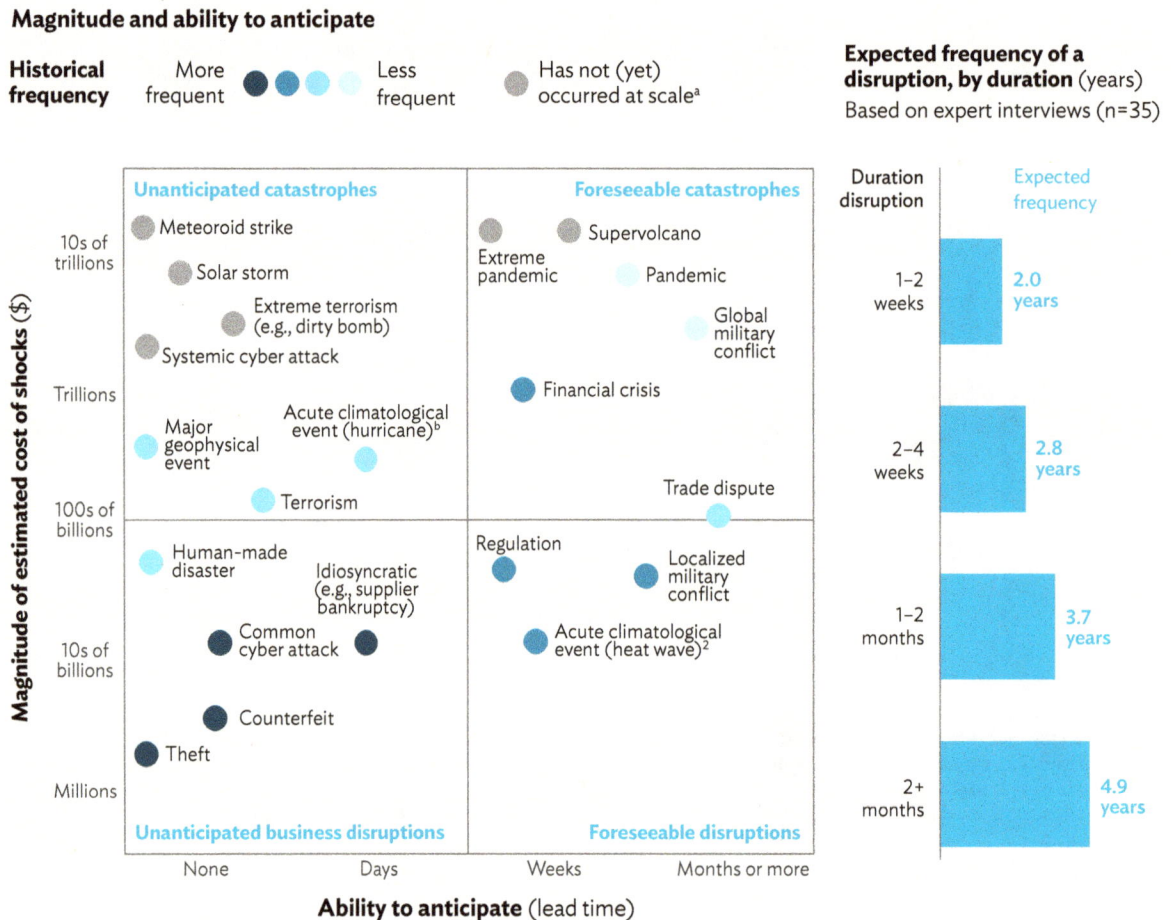

**Figure 31:** Profiling the Magnitude and Predictability of Disruptions

**Magnitude and ability to anticipate**

Historical frequency: More frequent ●●●● Less frequent | ● Has not (yet) occurred at scale[a]

**Expected frequency of a disruption, by duration** (years)
Based on expert interviews (n=35)

*Unanticipated catastrophes*
- Meteoroid strike
- Solar storm
- Extreme terrorism (e.g., dirty bomb)
- Systemic cyber attack
- Major geophysical event
- Acute climatological event (hurricane)[b]
- Terrorism

*Foreseeable catastrophes*
- Supervolcano
- Extreme pandemic
- Pandemic
- Global military conflict
- Financial crisis
- Trade dispute

*Unanticipated business disruptions*
- Human-made disaster
- Idiosyncratic (e.g., supplier bankruptcy)
- Common cyber attack
- Counterfeit
- Theft

*Foreseeable disruptions*
- Regulation
- Localized military conflict
- Acute climatological event (heat wave)[2]

Y-axis: Magnitude of estimated cost of shocks ($): 10s of trillions, Trillions, 100s of billions, 10s of billions, Millions

X-axis: Ability to anticipate (lead time): None, Days, Weeks, Months or more

| Duration disruption | Expected frequency |
| --- | --- |
| 1–2 weeks | 2.0 years |
| 2–4 weeks | 2.8 years |
| 1–2 months | 3.7 years |
| 2+ months | 4.9 years |

[a] Shocks that have not occurred either at scale (e.g., extreme terrorism, systematic cyber attack, solar storm) or in modern times (e.g., meteoroid strike)
[b] Based on experience to date, frequency, and/or severity of events could increase over time.
Source: S. Lund et al. 2020. Risk, Resilience, and Rebalancing in Global Value Chains. *McKinsey*. 6 August.

# 10.3. Risk Management and Crisis Learning

Faced with this new reality of more frequent and costly shocks, the Federation of European Risk Management Associations teamed up with McKinsey to survey more than 200 executives about the impact of the COVID-19 pandemic on corporate resilience[350] and their views on which dimensions of resilience[351] matter most for their strategy and operations (Figure 32).

---

[350] Federation of European Risk Management Associations and McKinsey, "The Role of Risk Management in Corporate Resilience," *Federation of European Risk Management Associations—FERMA* (blog), June 2021. https://www.ferma.eu/publication/the-role-of-risk-management-in-corporate-resilience/.; and McKinsey. 2022. From Risk Management to Strategic Resilience. March. https://www.mckinsey.com/capabilities/risk-and-resilience/our-insights/from-risk-management-to-strategic-resilience.

[351] The resilience dimensions included financial; operational; digital and technological; organizational; market position and innovation; reputation, brand, and customer; purpose and ESG capabilities; foresight; and disruption and crisis response.

## Figure 32: Assessing the Value of Resilience

**Resilience aspects reported as "very relevant" by sector** (% of respondents)

| Resilience dimensions | Advanced industries¹ | Finance | Consumer | GEM | Professional services | TMT | TLI | Health | Overall |
|---|---|---|---|---|---|---|---|---|---|
| Financial | 94 | 89 | 100 | 100 | 93 | 85 | 92 | 71 | 90 |
| Operational | 100 | 85 | 100 | 92 | 93 | 100 | 92 | 89 | 94 |
| Digital and technological | 87 | 89 | 92 | 95 | 100 | 100 | 100 | 100 | 95 |
| Organizational | 71 | 71 | 85 | 89 | 93 | 77 | 67 | 94 | 82 |
| Market position and innovation | 87 | 61 | 92 | 86 | 50 | 77 | 77 | 59 | 74 |
| Reputation, brand, and customer | 87 | 88 | 100 | 87 | 87 | 62 | 77 | 94 | 85 |
| Purpose and environmental, social, and governance | 71 | 74 | 92 | 92 | 69 | 62 | 77 | 88 | 78 |
| Foresight² | 67 | 71 | 62 | 69 | 50 | 58 | 46 | 65 | 61 |
| Disruption and crisis response | 80 | 74 | 92 | 76 | 53 | 69 | 100 | 78 | 78 |

*Industry sectors →*

Notes:
¹ Advanced industries includes advanced electronics, semiconductors, automotive and assembly, and aerospace and defense; Finance includes banking, insurance, and private equity; consumer includes consumer packaged goods; GEM includes basic materials, chemicals, and agriculture; Health includes health care, pharmacies, and social and public entities.
² Foresight refers to stress-testing to assess potential impact scenarios and simulated reactions on business and the capacity to identify resilience levers to reduce adverse effects.
Source: *McKinsey*. 2022. McKinsey on Risk. April.

Executives conceded that previous practices had focused on a small number of well-defined risks, especially financial risks. After the COVID-19 pandemic, this is being replaced by a much broader mandate of resiliency management:

- Nearly two-thirds of executives confirmed that resilience is central to their organizations' strategic process. Executive teams and risk functions are at the forefront of building a resilient organization. However, risk managers are not yet at the center of crisis resolution. This suggests that a better risk governance model is needed for efficient and effective decision-making and crisis management.
- Overall, the three dimensions of resilience viewed as most relevant by industry executives across all sectors are *digital and technological, operational, and financial* (Figure 32). The need for secure and flexible digital infrastructure and the close integration of digitalization with other resilience areas, such as working-from-home arrangements, is the new reality.

- Foresight capabilities (scenarios and stress-testing) emerged as one of the core areas for improvement to pressure-test strategies and business models, considering uncertainty and volatilities.
- Looking ahead, three-quarters of risk managers assign priority to initiatives that strengthen the risk culture and embed resilience firmly in the strategy process. Additional areas for improvement included the aggregation of risk-data, evidence-based reporting, and advanced foresight capabilities.

The importance assigned to digital resilience has far-reaching implications, both positive and negative. The lack of global standards for interoperability and the absence of legal and ethical accountability systems to protect people's rights and well-being against malicious use of data and technologies are obstacles to fully leverage these opportunities.[352] Although these challenges cut across different digital domains, the interaction between the various risks and their effects are often not considered but tackled in silos.

## 10.4. The Narrow View of Building Digital Resilience and Its Risks

The traditional approach is to frame digital resilience rather narrowly at individual, organizational, and national levels—an approach which is associated with a variety of shortcomings and risks.

Looking at it from the perspective of an individual, digital resilience is considered as an essential 21st century skill that centers on the ability of individuals to recognize and deal with the risks they encounter when they socialize, explore, or work online. The UK government's Digital Resilience Working Group, for instance, defines individual digital resilience as (i) understanding when a person is at risk online and can make informed decisions about the digital space one is in, (ii) knowing what to do to seek help from a range of appropriate sources, (iii) learning from experience and able to adapt future choices, and (iv) recovering when things go wrong online by receiving the appropriate level of support to aid recovery.[353]

Strategies for building individual digital resilience have included legislations that reduce potentially harmful content available online, as well as awareness-raising and educational activities on online safety and digital citizenship. However, these strategies are often implemented as ad hoc activities with limited support and guidance over time for individuals to recover and re-engage with digital opportunities. Moreover, research from the London School of Economics calls for a better understanding of individuals' interaction with the political, media, and economic environments that affect their digital resilience, such as polarization of society, low trust in news media, violence, and discrimination.[354] In addition, their social and cultural environments must be considered, including their education, health, and mental well-being.[355]

[352] S. Bashir et al. 2021. *The Converging Technology Revolution and Human Capital: Potential and Implications for South Asia.* Washington, DC: World Bank. https://openknowledge.worldbank.org/handle/10986/36156.

[353] UKCIS Digital Resilience Working Group. The Digital Resilience Framework. https://www.drwg.org.uk/the-framework.

[354] S. Banaji and R. Bhat. 2020. India: The Dissemination of Misinformation on WhatsApp Is Driving Vigilante Violence against Minorities, Minority Rights Group. 29 September. https://minorityrights.org/trends2020/india/.

[355] C. Moreno et al., 2020. How Mental Health Care Should Change as a Consequence of the COVID-19 Pandemic. *The Lancet. Psychiatry.* 7 (9): 813–24, https://doi.org/10.1016/S2215-0366(20)30307-2. According to the WHO, the Covid-19 pandemic has triggered a 25% increase in anxiety and depression worldwide.

At the organizational level, digital resilience is primarily concerned with safeguarding the digital network and data of the organization and preparing for a recovery from cybersecurity threats. Digital resilience typically is understood as the capability to anticipate, resist, and respond to disruptive cyber events and to recover from them within an acceptable timeframe. This narrow view of organizational digital resilience has gradually begun to include the continuity of business operations and increase in competitiveness through renewal and transformation. Strategies for organizational digital resilience include cybersecurity and data protection measures, disaster recovery and business continuity planning, crisis management, systems and data backup, and data analytics for modeling risk scenarios and early warning.

At the national level, digital resilience has been defined as the ability of the national ICT infrastructure to withstand and recover from external threats such as disasters, pandemics, IP theft, and cyber attacks. Some governments have classified digital resilience as part of critical infrastructure protection, reflecting the fact that other critical infrastructures (e.g., energy, transport, finance, education, health) are also increasingly dependent on digital systems.

Strategies for national IT infrastructure resilience include ensuring redundancy and diversity of network routes and equipment (e.g., deploying hybrid mesh and ring structure networks) and setting up backup services for emergency response efforts. Other strategies are related to the speed and scale at which services can be restored and recovered and the preparedness planning required.

Australia, for instance, has a critical infrastructure resilience strategy to safeguard the country's infrastructure from national security risks of sabotage and espionage. It involves identifying critical infrastructure, developing national security risk assessments and risk management strategies, and promoting compliance.[356] In the 2021 Dutch Digitalization Strategy,[357] one of its six priorities is digital resilience that focuses exclusively on cybersecurity, which involves promoting public–private partnerships, particularly with small and medium-sized enterprises (SMEs) and the high-tech sector.[358] In Malaysia, the National Cyber Security Agency was established with the objectives of securing and strengthening Malaysia's resilience in facing the threats of cyber attacks by collaborating with the nation's best experts and resources.[359]

Despite the increasing urgency for strengthening the digital resilience of individuals, organizations, and nations, research on translating these efforts into comprehensive strategies has been very limited up to now. Most public and private sector organizations struggle to define their resilience agenda and approach. Institutions are not yet fully prepared for the new reality, often reacting separately to each disruption, which is counterproductive.

---

[356] Australian Government, Department of Home Affairs . Critical Infrastructure Resilience Strategy. https://www.cisc.gov.au/what-is-the-cyber-and-infrastructure-security-centre/critical-infrastructure-resilience-strategy.

[357] Ministerie van Economische Zaken en Klimaat, "The Dutch Digitalisation Strategy 2021—About NL Digital—Nederland Digitaal," webpagina (Ministerie van Economische Zaken en Klimaat, June 22, 2021), https://www.nederlanddigitaal.nl/english/the-dutch-digitalisation-strategy-2021.

[358] In reviewing the strategy, the National Coordinator for Counterterrorism and Security observed that the Netherlands does not yet have digital resilience sufficiently under control due to (i) unauthorized access to information, particularly through espionage; (ii) inaccessibility of processes caused by sabotage and the use of ransomware; (iii) violation of the security of digital space; and (iv) large-scale outages. The Cyber Security Council concluded in its report *Integral approach to cyber resilience* that even in organizations that are part of vital processes, basic ICT and security hygiene is often not in order, so basic threats to their processes cannot be detected or deflected. K. Loohuis. 2021. The Netherlands Still Lacks Digital Resilience, Says Report. *ComputerWeekly*. 11 August. https://www.computerweekly.com/news/252505045/The-Netherlands-still-lacks-digital-resilience-says-report.

[359] National Cyber Security Agency. https://www.nacsa.gov.my.

Over the last decade, digital resilience strategies have remained focused on developing and protecting digital infrastructures, systems, data, and products. This approach has proven highly inadequate for responding to the COVID-19 pandemic and resulted in costly resilience failures. According to research by the World Economic Forum and McKinsey, workforce attrition during the pandemic may have shaved 3.6% off growth in some countries, while the impact caused by a lack of digital resilience, trust, and inclusion is estimated at 2%–2.5% of gross domestic product growth globally (Figure 33). On the flipside, it is estimated that success in reskilling and upskilling the labor force in the digitizing economy could increase growth by 4.5% annually by 2030.

The current resilience discussion is still characterized by differences in interpretation and vagueness on objectives, measurability, and areas for action. Few institutions have built sufficient strategic resilience, hampered by the additional cost of building redundancy, organizational forgetfulness, and a lack of a

### Figure 33: The Cost of Resilience Failures

**Estimates of impact on global annual GDP growth** (% change)

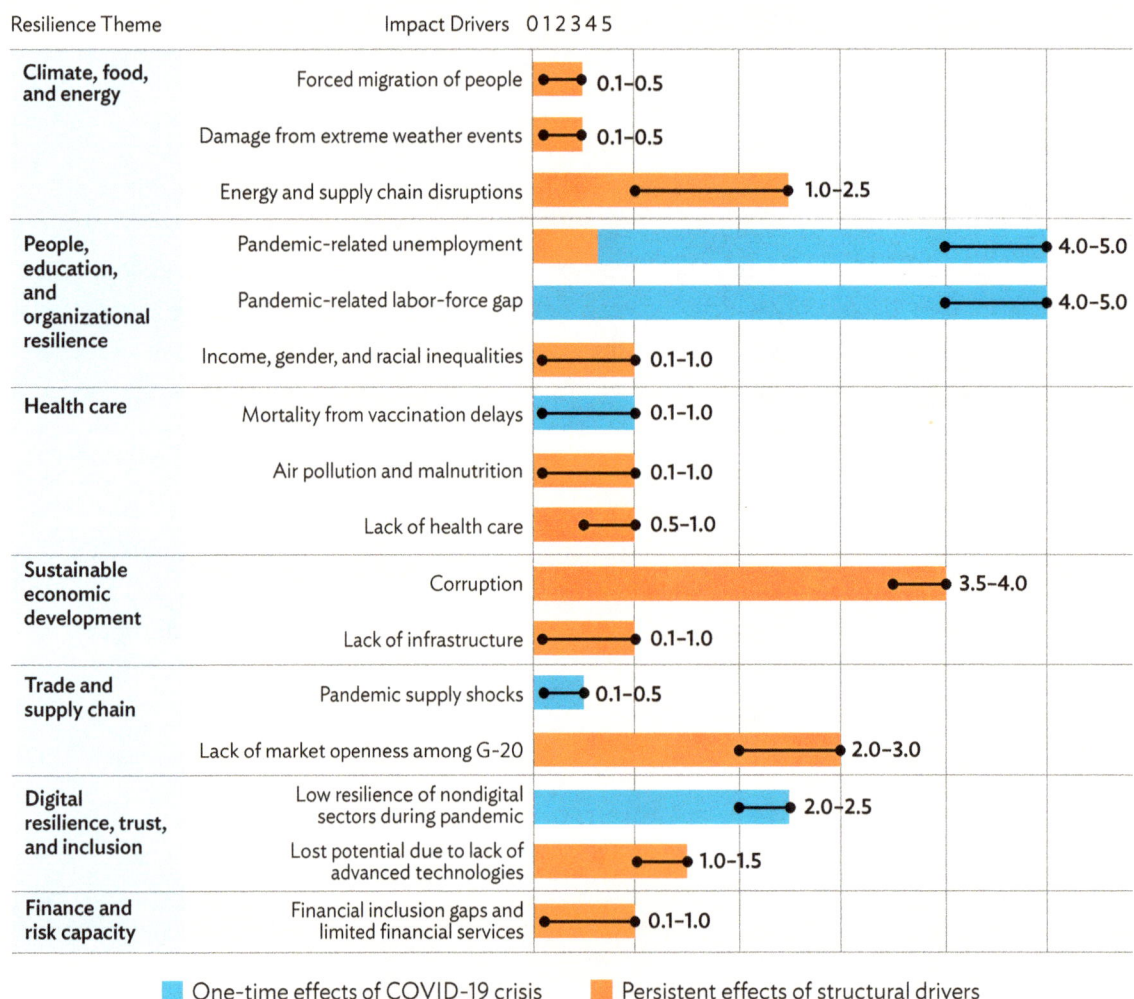

| Resilience Theme | Impact Drivers | Range |
|---|---|---|
| **Climate, food, and energy** | Forced migration of people | 0.1–0.5 |
| | Damage from extreme weather events | 0.1–0.5 |
| | Energy and supply chain disruptions | 1.0–2.5 |
| **People, education, and organizational resilience** | Pandemic-related unemployment | 4.0–5.0 |
| | Pandemic-related labor-force gap | 4.0–5.0 |
| | Income, gender, and racial inequalities | 0.1–1.0 |
| **Health care** | Mortality from vaccination delays | 0.1–1.0 |
| | Air pollution and malnutrition | 0.1–1.0 |
| | Lack of health care | 0.5–1.0 |
| **Sustainable economic development** | Corruption | 3.5–4.0 |
| | Lack of infrastructure | 0.1–1.0 |
| **Trade and supply chain** | Pandemic supply shocks | 0.1–0.5 |
| | Lack of market openness among G-20 | 2.0–3.0 |
| **Digital resilience, trust, and inclusion** | Low resilience of nondigital sectors during pandemic | 2.0–2.5 |
| | Lost potential due to lack of advanced technologies | 1.0–1.5 |
| **Finance and risk capacity** | Financial inclusion gaps and limited financial services | 0.1–1.0 |

■ One-time effects of COVID-19 crisis    ■ Persistent effects of structural drivers

COVID-19 = coronavirus disease, GDP = gross domestic product.
Source: B. Brende and B. Sternfels. 2020. Resilience for sustainable, inclusive growth. *McKinsey*. 7 June.

universal metric to measure and evaluate the efficacy of investments in resilience. By restricting the resilience agenda just to the physical and technical aspects, opportunities offered by digital technologies are at risk of being overlooked. Instead, the concept of digital resilience should be expanded to consider ways to reduce vulnerabilities and increase opportunities.

Digital resilience requires multi-stakeholder engagement to identify and address the risks of digital technologies and establish mechanisms for joint action that foster inclusiveness and digital rights, especially for vulnerable and marginalized groups.[360] It is important to gain a better understanding of people's access to and use of digital technologies, including their behaviors and perceptions online. However, the current narrow view of digital resilience hinders a full understanding of the risks and ignores the human aspects, particularly the diverse needs, vulnerabilities, capacities, and rights of different communities and individuals. The failure to include marginalized and vulnerable groups in digital resilience efforts also risks amplifying the digital divide as they are left further behind.

At the organizational level, it is no longer sufficient for organizations to merely safeguard their own digital systems and data. Organizations need to engage beyond their own digital systems and data to safeguard their supply chains and the communities they serve. A case in point are SMEs, whose challenges during the COVID-19 pandemic were not just limited to digital access and protection from cybersecurity threats. They also had to adapt to supply chain disruptions, find innovative solutions to meet changing market demands, comply with new government regulations, shift their work environments to digital platforms, and ensure the safety and well-being of employees and their families.

To address these complex challenges, policymakers and governments need to collaborate across agency silos, adopt favorable regulations and incentives for building digital resilience, incorporate social protection measures, and increase the availability of finance and training opportunities for SMEs. The European Commission has perhaps gone furthest in this regard by offering guidance for embedding resilience in all future policymaking. The EU's first Strategic Foresight Report defines digital resilience as: "Ensuring that the way we live, work, learn, interact and think in this digital age preserves and enhances human dignity, freedom, equality, security, democracy, and other European fundamental rights and values."[361] It is an attempt to view resilience holistically by focusing on four interrelated dimensions of resilience: digital, social, and economic, geopolitical, and green.

---

[360] UN Secretary-General High Level Panel. The Age of Digital Interdependence. https://digitallibrary.un.org/record/3865925.
[361] European Commission. 2020 Strategic Foresight Report. https://ec.europa.eu/info/strategy/strategic-planning/strategic-foresight/2020-strategic-foresight-report_en.

## 10.5. From Crisis Response to a Holistic Digital Resilience Strategy

The COVID-19 pandemic has provided an opportunity to rethink and reformulate strategies for digital resilience toward current and future threats and crises.

The Republic of Korea's comparatively successful pandemic response has been attributed to its holistic approach involving close collaboration between the government, the scientific community, and the private sector; the release of open data; strong protection of data privacy and security; and transparency in reporting and tracking.[362] Similarly, New Zealand placed an emphasis on cooperation, government transparency, quick decisive action based on scientific data, and the holistic people-centered deployment of interventions.[363] The Institute for Health Metrics and Evaluation published an analysis indicating that countries that fared the best with their COVID-19 responses were those where the trust in government was highest.[364]

This broader concept of digital resilience encompasses the physical, economic, social, political, and environmental vulnerabilities and capabilities of nations, organizations, and individuals and takes account of their impact on each other. A holistic approach to digital resilience would not only enhance access to and security of digital systems, but also empower people and communities to use digital technologies to pursue economic and social opportunities in a safe and secure manner.

Digital technologies can help strengthen resilience through enhanced data collection, processing, diffusion, and delivery capabilities. Companies and entrepreneurs who are able to give up a narrow focus on risk, controls, and governance for a longer-term holistic strategic view will be able to turn resilience into a competitive advantage in times of disruption.

A comprehensive digital resilience agenda should include the following:

1. An IT infrastructure that can ensure redundancy and diversity of network routes and equipment and increase the speed and scale at which resources can be mobilized and accessed for restoration of services when attacked or hit by disasters.
2. Cybersecurity, digital privacy, and trust-building measures to protect and assure national, organizational, and individual resilience. Adopting security-by-design and privacy-by-design principles and protecting vulnerable groups need to be elevated as a collective responsibility.
3. Deployment of digital technologies to collect, process, map, and visualize data on disaster risk which serve as mission-critical platforms for multi-stakeholder engagements and early warning systems.

---

[362] D. Lee, K. Heo, and Y. Seo. 2020. COVID-19 in South Korea: Lessons for Developing Countries. *World Development*. 135: 105057, https://doi.org/10.1016/j.worlddev.2020.105057.

[363] M.G. Baker. 2020. Successful Elimination of Covid-19 Transmission in New Zealand. NEJM. 20 August. https://www.nejm.org/doi/full/10.1056/NEJMc2025203.

[364] Institute for Health Metrics and Evaluation. 2022. Lack of trust has helped fuel the COVID-19 pandemic, new study shows. News release. 1 February. https://www.healthdata.org/news-release/lack-trust-has-helped-fuel-covid-19-pandemic-new-study-shows.

4. Digital technologies that contribute to more resilient food security and supply chains, complemented with the expansion of energy access and IT infrastructure, education and skills development, and increased investment in rural areas through integrated approach.

5. Digital innovations in the health sector need to be balanced to ensure inclusiveness and protect individual rights to privacy and security, including deployments of AI and Big Data.

6. Comprehensive measures to protect mental wellness for all are needed. Mental distress, violence against women and children (online and offline), and substance abuse are worsening during disasters and crises, yet are often sidelined as resources are diverted to emergency response.

7. A re-formulation of education, training, and lifelong learning strategies to help ensure holistic upskilling and reskilling, which is critical for building digital resilience. This involves enabling all citizens to learn online and co-create solutions while protecting their basic rights and safety. Digital financial services could, for example, serve as an entry point to learn about digital identifiers, data protection, and trust.

8. Boosting of network capacity, developing appropriate skills, and putting in place privacy and security measures for telework as well as for e-ecommerce, which are necessary for maintaining productivity and business continuity. Empowering marginalized and vulnerable groups should be a priority as they typically earn less, have less-secure jobs (i.e., gig work), and are less likely to have assets to fall back on during crises.

9. Empowerment of local entrepreneurs, social enterprises, and SMEs to tap into their innovative potential to serve their local communities, particularly during crisis situations.

10. Development of skills to co-create locally driven content and applications that ensure access and inclusion can help make digital infrastructure more resilient.

11. Policies and investments that promote universal meaningful connectivity, expanding access to unserved and underserved areas that may not be commercially viable and making connectivity affordable.

12. Actions by the fintech industry addressing the significant digital risks that users are exposed to (e.g. privacy and security risks, lack of digital literacy, and exclusion of women and offline communities).

13. Timely and accurate data collection by governments and open access to disaggregated data sets which quality statistics and data insights rely on. Only then can risks be better understood, signals and trends analyzed, and interventions designed and delivered to those in need in a targeted manner.

14. Access to timely and credible information plays an important role during crisis situations. Hence, digital technologies which can contribute to the identification and prevention of misinformation deserve more attention.

**Cybersecurity infrastructure.**
Important to protect digital assets
and to ensure a secure digital
landscape (photo by ADB).

# APPENDIX 1
# CYBERSECURITY GLOSSARY[1]

**Advanced Persistent Threat (APT)** is an adversary with sophisticated levels of expertise and significant resources, allowing it, through the use of multiple different attack vectors (e.g., cyber, physical, and deception), to establish footholds within the information technology (IT) infrastructure of organizations for purposes of exfiltrating information and/or to undermine or impede critical aspects of a mission, program, or organization, or place itself in a position to do so in the future.

**Attack Vectors.** Hackers use attack vectors to gain access over a network or a computer to infect the system with malware or to harvest data. Common attack vectors include man-in-the-middle, drive-by, Structured Query Language (SQL) injection, and zero-day attacks.

**Bots** are automated applications or scripts designed to perform repetitive tasks without requiring human input. Bots can be used to maliciously distribute spam, conduct distributed denial-of-services (DDoS) attacks, operate as malware command-and-control infrastructure, or flood public forums with fraudulent commentary to propagate a specific message.

**Computer Emergency Response Team (CERT)** is an expert group that handles computer security incidents. Alternative names for such groups include **computer security incident response team (CSIRT)**.

**Critical Infrastructure** describes the physical and cyber systems and assets that are so vital to a country that their incapacity or destruction would have a debilitating impact on its physical or economic security or public safety.

**Cyber Attack** is an action taken to harm a system or disrupt normal operations by exploiting vulnerabilities using various techniques and tools.

**Cybercrime-as-a-Service (CaaS)** involves the productizing of malware and the on-demand purchasing and selling of cybercrime services.

**Cyber Insurance** is designed to help businesses hedge against the potentially devastating effects of cybercrimes such as malware, ransomware, DDoS attacks, or any other method used to compromise a network and sensitive data. Demand for cyber insurance has grown rapidly in recent years.

---

[1]    USAID. Cybersecurity Primer and other sources.

**Cyber Kill Chain** is a framework to identify and prevent cyber intrusions. The model identifies the steps the adversaries must complete to achieve their objective. It strengthens a cyber analyst's understanding of an adversary's tactics, techniques, and procedures.

**Cyber Resiliency** is the ability to prepare for and adapt to changing conditions and withstand and recover rapidly from disruptions. Resilience includes the ability to withstand and recover from deliberate attacks, accidents, or naturally occurring threats or incidents.

**Cyber Risks** involve the potential for financial loss, disruption, or damage to the reputation of an individual, organization, or government from failure, unauthorized or erroneous use, or other malicious exploitation of its information systems.

**Cyber Security** is the management of cyber risks and vulnerabilities to mitigate cyber threats.

**Cyber Threat** is an action that takes advantage of security weaknesses in a system and has negative impact on it. Threats can originate from two primary sources: humans and nature. Few safeguards can be implemented against disasters. Human threats consist of internal (someone has authorized access) or external threats (individuals or organizations working outside the network).

**Cyber Vulnerabilities** are specific weaknesses in a computer system or online network such as a lack of encryption or poorly designed firewalls that allow an intruder to execute commands, access unauthorized data, and/or conduct denial-of-service attacks. It can also be a deficit in capacity or skills of people to protect those systems or networks.

**Digital Ecosystem** comprises stakeholders, systems, and an enabling environment that, together, empowers people and communities to use digital technology to access services, engage with each other, and pursue economic opportunities. It is organized around three overlapping pillars: digital infrastructure and adoption; digital society, rights, and governance; and the digital economy. It also encompasses four crosscutting topics: inclusion, cybersecurity, emerging technologies, and geopolitical positioning.

**Digital Literacy** is the ability to access, manage, understand, integrate, communicate, evaluate, and create information safely and appropriately through digital devices and networked technologies for participation in economic and social life.

**Digital Trust** is created when users have confidence in an online system, network, or technology, and trust that their data and privacy are being protected when using them.

**Internet of Things (IoT)** devices pose a potential cybersecurity risk because they share a lot of information online and do not necessarily adhere to the highest levels of cybersecurity protections.

**Insider Threat** refers to a person(s) within an organization who poses a potential risk through knowingly or unknowingly violating security policies.

**Man-in-the-Middle Attack** is a form of cyber eavesdropping in which malicious actors insert themselves into a conversation between two parties and intercept data through a compromised but trusted system. The targets are often intellectual property or fiduciary information.

**Multi-factor Authentication** is a security process that requires more than one method of authentication from independent sources to verify a user's identity.

**Ransomware** is a type of malware that accesses a victim's files, locks and encrypts them, and then demands the victim pay a ransom to get them back.

**Rootkit** is a collection of computer software, typically malicious, designed to enable access to a computer or an area of its software that is not otherwise allowed and often masks its existence or the existence of other software. The term rootkit is a compound of "root" and the word "kit."

**Spoofing** is the act of disguising a communication from an unknown source as being from a known, trusted source.

**Unified Extensible Firmware Interface (UEFI)** is a publicly available specification that defines a software interface between an operating system and platform firmware.

**Zero Day Attack or Exploit** is an advanced cyber attack that happens once a flaw, or software/hardware vulnerability, is exploited and attackers release malware before a developer has an opportunity to create a patch to fix the vulnerability.

# APPENDIX 2
# KEY CYBER ACTORS AND INSTITUTIONS

**African Union (AU).** Africa faces a growing array of cyber threats from espionage, critical infrastructure sabotage, and organized crime. AU is developing a continental cybersecurity strategy through its recently established African Cyber Experts (ACE) group and is partnering with the Global Forum on Cyber Expertise (GFCE) to support cyber capacity building.[2] The AU has organized cybersecurity training (on national cybersecurity strategy, CERT/CSIRT; modern cyber legislation) with external partners in more than 50 member states.

**Asian Infrastructure Investment Bank (AIIB).** As part of its "Infrastructure of Tomorrow" (i4t) program, AIIB launched a Digital Infrastructure Sector Strategy in 2020. While there is no systemic approach to cybersecurity or cyber resilience as part of its lending operations, AIIB has identified cybersecurity as one of the main regulatory risks to investing in infrastructure, which is now being incorporated as a component of some of their projects. AIIB has built an in-house capacity to ensure that its digital infrastructure projects comply with specific country regulations and laws and are based on regulatory risk analysis that balances out data privacy risk with reputational risk.

**Australia Department of Foreign Affairs and Trade (DFAT).** In 2021, DFAT launched an expanded International Cyber and Critical Technology Engagement Strategy,[3] with a strong focus to support countries in the Indo-Pacific region in their cyber capabilities and resilience efforts (including fighting cybercrime, improving online safety, and countering disinformation). Australia's strategy incorporates cybersecurity de-risking measures similar to environmental and social safeguards.

**CREST** is an international not-for-profit accreditation and certification body that represents the global cybersecurity industry. CREST awards internationally recognized accreditations to around 300 member companies and over 3,500 certified individuals who are providing vulnerability assessments, penetration testing, cyber incident response, and threat intelligence. In collaboration with industry and governments, CREST has built a framework and platform for measuring the capability of cybersecurity companies and their workforce. With Gates Foundation funding, CREST has developed an open-source Cybersecurity Incidence Response Maturity Model to measure the maturity and financial inclusion of the cybersecurity ecosystem.[4]

---

[2] A. Ajijola nd N. Allen. 2022. African Lessons in Cyber Strategy. *Africa Center for Strategic Studies.* 8 March. https://africacenter. org/spotlight/african-lessons-in-cyber-strategy/.

[3] Australian Government. 2021 International Cyber and Critical Technology Engagement Strategy. https://www.internationalcybertech.gov.au/our-work.

[4] CREST. Cyber Security Incident Response Maturity Assessment. https://www.crest-approved.org/approved-services/cyber-security-incident-response-maturity-assessment/.

**Cybersecurity Industry Associations** is a listing of ~100 cybersecurity industry associations can be found on the website of the Cybercrime Magazine.

**Cybil** is a GFCE-managed knowledge hub of international cyber capacity building, including tools, publications, and overview of activities and organizations on cyber capacity building globally. Highly recommended.

**Estonia** has developed significant cyber expertise[5] and regularly invests in cyber capacity-building projects, especially with Eastern European partners (Ukraine, Georgia), focused on supporting government's digital infrastructure, services, and cybersecurity capacity; updating legislation; improving system interoperability; and strengthening citizens' trust in ICTs. Estonia's e-Governance Academy maintains the National Cyber Security Index (NCSI), a global live benchmarking index, which measures the preparedness of countries to prevent cyber threats (denial of e-services, data integrity breach, data confidentiality breach) and manage cyber incidents. The NCSI is also a tool for national cyber security capacity building.[6]

**European Investment Bank (EIB).** EIB-financed projects must comply with high technical environmental, social, and governance (ESG) standards. The European Investment Advisory Hub (a joint advisory initiative of EIB and European Community) and the European Cyber Security Organization are set to create a European Cybersecurity Investment Platform to attract €1 billion investment over the next 5 years for European cybersecurity.[7]

**European Bank for Reconstruction and Development (EBRD).** The EBRD has begun incorporating cybersecurity and digital resilience safeguards into its lending operations, conducting digital risk assessments for sector-specific projects that involve digital technologies, with closer scrutiny applied to projects that involve personally identifiable information. The EBRD recently adopted its first Digital Approach to accelerating digital transition, which "sets out a comprehensive framework on how the Bank will use investment, policy engagement, and advisory services to support the digital transition in the economies where it invests ... and aims to mainstream technology throughout the Bank's activities." The strategy expressly recognizes cybersecurity and the protection of data privacy as essential parts of the digital transition and states that the EBRD will develop a crosscutting approach to cyber resilience to protect itself, its clients, and the economies in which it operates.

**Gates Foundation (BMFG).** The Gates Foundation recognizes that cybersecurity is fundamental to ensuring digital financial inclusion and developing a more secure world. The foundation supports efforts to create Digital Public Goods, including tools to measure the cyber maturity of the financial sector in developing countries (CREST), threat models for the financial sector in low-income countries (MITRE Engenuity), and networks of universities focused on improving cybersecurity of financial systems.

**Global Forum on Cyber Expertise (GFCE)** is a global coordinating platform of countries, international organizations, and the private sector used to exchange best practices and expertise in cyber capacity building.

---

5   K. Kohler. 2020. *Estonia's National Cybersecurity and Cyberdefense Posture: Policy and Organizations.* Zürich: ETH Zürich Center for Security Studies. https://doi.org/10.3929/ETHZ-B-000438276.
6   National Cyber Security Index. https://ncsi.ega.ee/ncsi-index/.
7   European Cyber Security Organisation. European Cybersecurity Investment Platform. https://ecs-org.eu/initiatives/european-cybersecurity-investment-platform-ecip.

**Inter-American Development Bank (IDB).** The IDB provides funding for digital transformation strategies and cybersecurity initiatives. A multiyear project to support Chile's cybersecurity readiness aims to build resilience to digital threats, develop personnel, and strengthen critical infrastructure and government institutions. The IDB also supports the cybersecurity agenda in key sectors such as health, energy, finance, water, and transportation, as illustrated by the recent release of a guide to help cities in Latin America and the Caribbean to be prepared for malicious attacks.[8]

**International Bank for Reconstruction and Development (World Bank).** A majority of World Bank projects have ICT as a core component, and each of its 14 technical Global Practices is introducing digital technologies and beginning to incorporate a cybersecurity subcomponent as part of digital economy projects. The World Bank has also been involved in the Global Cybersecurity Capacity Program since 2016. The Digital Development Global Practice offers cybersecurity technical assistance in five areas: (i) technical and legal support (including policy dialogues, strategy development, benchmarking); (ii) diagnostics (national cybersecurity assessment, best practices to mitigate cyber risk to the financial sector); (iii) institutional and governance framework (cybersecurity regulation for SMEs); (iv) technical support (strengthening national CERTs/SOCs, procurement, threat intelligence gathering, support for critical infrastructure agencies, and agencies handling biometric data); and (v) digital skills development and cybersecurity training. Under a grant from the Republic of Korea–World Bank Partnership Facility, the World Bank has also developed a specific tool kit to combat cybercrime.[9] In August 2021, it launched a Cybersecurity Multi-Donor Trust Fund with support from the Netherlands, Estonia, Germany, and Japan to advance the cybersecurity development agenda.

**International Finance Corporation (IFC).** The IFC has incorporated de-risking processes in its digital investment strategy and developed a risk management framework for specific industry sectors to help assess risk (High–Medium–Low). The sectors considered to have higher exposure to cyber risks (e.g., energy, logistics) receive additional scrutiny versus sectors considered lower risk (e.g., finance, telecom) because of existing industry frameworks, best practices, regulations, and experience.

**International Computer Security Incident Response Teams (CSIRTs).** Carnegie Mellon University maintains a list of national CSIRTs across the world.

**International Telecommunications Union (ITU).** The ITU Cybersecurity program involves five areas: (i) develop effective national cybersecurity strategies (NCS), as outlined in the updated 2021 NCS guide; (ii) develop national CSIRTs and conduct Cyber Drills; (iii) adopt appropriate cybersecurity legislation; (iv) promote inclusivity; and (v) combat spam.

**Islamic Development Bank (IsDB).** In 2019, IsDB adopted an ICT Policy that includes cybersecurity as an enabler in the context of corporate governance and access to e-government services.

---

8    L. Cotino and M. Sanchez. 2021. *A Cybersecurity Guide for Smart Cities*. London: Inter-American Development Bank. https://publications.iadb.org/publications/english/document/A-Cybersecurity-Guide-for-Smart-Cities.pdf.

9    World Bank and United Nations. 2017. *Combatting Cybercrime*: Tools and Capacity Building for Emerging Economies. Washington, DC: World Bank. https://doi.org/10.1596/30306.

**Israel's Agency for International Development Cooperation (MASHAV)** and **Israel National Cyber Directorate (INCD).** MASHAV activities are focused on capacity building and the transfer of Israeli know-how and innovative technologies, like cybersecurity, to regions that are of strategic interest to Israel (e.g., Central and South Asia, LAC, MNA). INCD—the country's center of cyber expertise—delivers capacity-building programs, which tend to be focused on securing critical digital infrastructure, helping establish national CERTs/SOCs, and developing core cyber capacity. Israel has attracted one-fifth of global private cybersecurity investment since 2019.[10]

**Japanese International Cooperation Agency (JICA).** JICA focuses its cybersecurity capacity-building initiatives through university partnerships to develop indigenous capacity. In Indonesia, for example, JICA is co-creating and implementing a masters-level program on cybersecurity.

**MITRE Engenuity,** a Tech Foundation for Public Good, is developing (with support by the Gates Foundation) a cyber threat-based risk model for digital financial mobile services in Africa and India. The model incorporates MITRE's threat-informed defense approach, attacker methods, and technology-specific vulnerabilities relating to mobile money applications.

**Norwegian Institute of International Affairs (NUPI).** NUPI's Centre for Digitalization and Cybersecurity Studies is seeking to bridge the gap between the technical community and the policy world with research focusing on the political dimension of cybersecurity and its role in international relations, global governance of cyberspace, capacity building, and development.

**Organization of American States (OAS).** The OAS's Cyber Security Program is focused on building cybersecurity capacity in its 35 member states through development of (i) national and regional cybersecurity strategies and legal frameworks; (ii) CSIRTs, protection of critical infrastructures, improved ability to monitor and respond to cyber incidents; and (iii) tool kits and policy reports. The OAS has also set up an Inter-American Cooperation Portal and Working Group on Cybercrime.

**Organisation for Economic Co-operation and Development (OECD).** The OECD has long been supporting cooperation on the management of digital security risk to economic and social prosperity[11] and is providing a forum for stakeholders to develop digital security policies that build trust in the global digital environment. The OECD launched a Global Forum on Digital Security for Prosperity in 2018 to champion multi-stakeholder dialogue and facilitate the convergence of views.

**Organization for Security and Co-operation in Europe (OSCE).** The OSCE has a mandate to support its 57 participating states in enhancing their criminal justice response to cybercrime while upholding human rights, fundamental freedoms, and the rule of law. The focus is on implementing confidence-building measures and cyber diplomacy.

---

[10] L. Tabansky. 2022. How Israel Became a Top Cyber Power. *The National Interest.* 21 March. https://nationalinterest.org/blog/techland-when-great-power-competition-meets-digital-world/how-israel-became-top-cyber-power.

[11] Organisation for Economic Co-operation and Development. 2015. *Digital Security Risk Management for Economic and Social Prosperity.* Paris https://www.oecd-ilibrary.org/science-and-technology/digital-security-risk-management-for-economic-and-social-prosperity_9789264245471-en.

**Singapore's Cyber Security Agency (CSA).** In May 2022, CSA and the Nanyang Technological University launched the National Integrated Centre for Evaluation (NiCE). The joint center serves as a one-stop facility for cybersecurity evaluation and certification and boost Singapore's branding as a cybersecurity hub.

**UN Executive Office of the Secretary-General (EOSG).** The UN Secretary-General has prioritized the use of digital technologies to strengthen human rights, advance peace, and improve all lives as part of the UN Sustainable Development Goals. Two key reports, *The High-Level Report on Digital Cooperation* and the *Data Strategy of the UN SG for Action*, highlighted the risks of key technologies that do not meet users' needs, loose trust by mismanaging cybersecurity and privacy, or lock users in inflexible "one-size-fits-all" systems.

**UN Office of the Secretary-General's Envoy on Technology (OSET).** OSET, created in 2021, has the mandate to coordinate the implementation of the 2020 UN Roadmap for Digital Cooperation, seizing on the opportunities while mitigating the risks. The office also launched a new Joint Facility for Global Digital Capacity (jointly with ITU and UNDP). The UN Shared Cyber Hub brings together the cyber focal points of over 20 UN bodies to ensure they speak with one voice on cybersecurity and share lessons learned.

**UN Conference on Trade and Development (UNCTAD).** The UNCTAD Global Cyberlaw Tracker offers a global mapping of cyberlaws. It tracks the state of e-commerce legislation, consumer protection, data protection/privacy, and cybercrime adoption in the 194 UNCTAD member states. While 156 countries (80%) have enacted cybercrime legislation, the evolving cybercrime landscape and resulting skills gaps are a significant challenge for law enforcement agencies, especially for cross-border enforcement.

**UN Development Programme (UNDP).** UNDP has developed a Whole-of-Society Digital Transformation framework as an overarching reference model as well as a Digital Readiness Assessment (DRA) tool to identify digital strengths and weaknesses. Other initiatives include a Misinformation project, and an AI Readiness project.

**UN Office on Drugs and Crime (UNODC).** UNODC is dedicated to strengthening member states' capacities to confront threats from transnational crime, terrorism, corruption, and drug trafficking. UNODC's Global Program on Cybercrime assists developing countries in their efforts to prevent and combat cyber-related crimes through law enforcement capacity building and technical assistance.

**UK Foreign, Commonwealth & Development Office (FCDO).** FCDO's Cyber Policy Department is addressing cybersecurity threats, building resilience to cybersecurity attacks, and promoting trusted and secure technology across the world, working with other UK agencies such as the National Cyber Security Centre. FCDO is embedding digital/cyber de-risking mechanisms in both its cybersecurity and digital development assistance program. Its Digital Access Program (which includes Indonesia) emphasizes digital inclusion, the importance of managing digital risk, and growing the local digital economy.

**USAID.** In April 2017, USAID released its first-ever Digital Strategy, which identified cybersecurity as a new focus area and outlined ways to improve development outcomes through the use of digital technologies.[12] USAID has also released a Digital Ecosystem Framework, composed of three pillars (Digital Infrastructure and Adoption; Digital Society, Rights and Governance; and Digital Economy) and four crosscutting topics (Inclusion, Cybersecurity, Emerging Technologies, and Geopolitical Positioning). The Digital Strategy informs the work of the newly established Cybersecurity Team. The team's newly released Cybersecurity Primer advocates for cybersecurity and digital resilience safeguards to become a first-order strategic and operational priority throughout USAID's programming cycle.[13] In 2019, USAID also launched the Digital APEX program to bridge the development and cybersecurity communities and offer a pool of pre-approved, US-based cybersecurity experts and companies that can assist beneficiaries in the event of a significant cyber event or help reduce digital vulnerabilities (e.g., security assessments, procurement, network monitoring, incident response).

---

[12]   USAID. Digital Strategy. https://www.usaid.gov/digital-strategy.
[13]   USAID. Cybersecurity Primer. https://www.usaid.gov/digital-development/usaid-cybersecurity-primer.

# APPENDIX 3
# CYBER SECURITY RESOURCES

*ASEAN Cyberthreat Assessment 2021 (Interpol, 2021).* The report outlines how cybercrime's upward trend is set to rise exponentially, with highly organized cybercriminals sharing resources and expertise to their advantage. It provides strategies for tackling cyberthreats and describes the essential collaboration on intelligence sharing and expertise between law enforcement agencies and the private sector.

*Considerations for Managing Internet of Things (IoT) Cybersecurity and Privacy Risks (NIST, 2019).* IoT is a rapidly evolving and expanding collection of diverse technologies that interact with the physical world. This publication identifies high-level considerations that may affect the management of cybersecurity and privacy risks for IoT devices as compared to conventional IT devices.

*CrowdStrike Global Threat Reports (CrowdStrike, 2021, 2022).* The report features analysis from the CrowdStrike Threat Intelligence team and highlights the most significant global events and trends in the past year. The report covers real-world scenarios and observes trends in attackers' evolving tactics, techniques, and procedures, and offers practical recommendations to protect and counteract.

*Cyber Incident Management in Low-Income Countries: Part 1 and Part 2* (GFCE, Global Affairs Canada, Africa CERT, 2022). This report discusses the findings and recommendations of the Cyber Incident Management in Low-Income Countries project funded by Global Affairs Canada. The project aims to create a tailorable guide for low-income countries to develop or improve their computer security incident response team (CSIRT) capabilities in an affordable way to respond to the evolving cyber threat environment effectively.

*Cyber-risk Oversight Handbook for Corporate Boards (Organization of American States, 2019).* Boards of directors must take a leading role in the oversight of the safety of their company's cyber systems. However, this study from the Organization of American States and the InterAmerican Development Bank found corporate boards in Latin America generally have low or medium levels of maturity and knowledge related to cybersecurity. Consequently, they may lack awareness of how cyber threats might specifically affect their organizations.

*Cybersecurity Capacity Maturity Model for Nations (CMM) (Global Cyber Security Capacity Centre (GCSCC), University of Oxford, 2021).* The *Cybersecurity Capacity Maturity Model for Nations* facilitates the assessment of a country's cybersecurity capacity maturity. Developed by the GCSCC in consultation with over 200 international experts drawn from governments, international organizations, academia, public and private sectors, and civil society, the CMM reviews cybersecurity capacity across five dimensions: (i) Developing cybersecurity policy and strategy, (ii) Encouraging responsible cybersecurity culture within society, (iii) Building cybersecurity knowledge and capabilities, (iv) Creating effective legal and regulatory frameworks, and (v) Controlling risks through standards and technologies. The CMM allows the benchmarking of national cybersecurity capacity.

*Cybersecurity Skills Development in the EU* (ENISA 2020). This report argues that many of the current issues in cybersecurity education could be lessened by redesigning educational and training pathways which students should possess upon graduation and after entering the labor market. Four countries—Australia, France, the United Kingdom, and the United States—have attempted to rethink cybersecurity degrees using certifications. ENISA has created the Cybersecurity Higher Education Database, which is an interactive database of cybersecurity degrees.

*Cybersecurity in Pacific Island Nations* (GCSCC Oxford University; Oceania Cyber Security Centre; ITU, 2020). While the detailed reports are confidential, general learnings and themes emerge from the work completed in Samoa, Tonga, Vanuatu, Papua New Guinea, and Kiribati. Cybersecurity is seen in a very wide sense. Particularly important topics are malicious use of social media, fake news, deception, and fake accounts. One main issue is the lack of qualified people to support governments, companies, educational institutions, and the public in general.

*Cybersecurity Survival Guide, 6th Edition* (Palo Alto Networks, 2022). New methods of processing and securing network traffic are being used to provide visibility and control over traffic, applications, and threats. Enterprise security deals with threat protection for large and complex organizations, while cybersecurity scales the vast landscape of the internet riddled with vulnerabilities and viruses.

*Cybersecurity in Working from Home: An Exploratory Study* (University of Oxford, 2021). This paper presents the findings of an exploratory study of the implications of a shift to working from home in the context of the coronavirus disease (COVID-19) pandemic.

*Cybersecurity Toolkit for Small Business* (Global Cyber Alliance, 2021). The tool kit is an online resource for businesses to assess their security posture, implement free tools, and find practical tips and free resources to help improve cybersecurity readiness and response.

*Cyber Peace Builders* is a network of corporate volunteers providing free assistance to nongovernment organizations (NGOs) protecting vulnerable populations anywhere in the world. The Cyber Peace Builders program provides free pre- and post-incident assistance to NGOs in critical civilian sectors, such as health care, water and sanitation, food and agriculture, energy, information, etc. Pre-incident assistance includes provision of security assessments, awareness-raising and incident-response capability building, and training NGOs so they know how to respond and who to contact. Post-incident assistance revolves around security hardening to avoid reinfection.

*Cyber Strategy Development & Implementation (CSDI) Framework* (MITRE, 2020). MITRE's CSDI framework draws on best practices of 18 US, international, and industry models. It uses a combination of design thinking, threat/opportunity/resources contextualization, and eight key cyber capacity areas to assess cyber needs and threats, developing risk-informed strategic goals and implementing them in a multi-stakeholder environment.

*Cyber Threat Intelligence Sharing* (UK Foreign, Commonwealth and Development Office, Citibank, 2021). This document explains the sharing of information on cyber security domestically, regionally, and globally, and across sectors.

*Darknet Cybercrime Threats to Southeast Asia* (United Nations Office on Drugs and Crime (UNODC, 2021). This report assesses the dark web from user, criminal, and law enforcement perspectives with a particular focus on cyber criminality targeted at Southeast Asian countries.

*Future Series: Cybersecurity, Emerging Technology and Systemic Risk* (World Economic Forum, University of Oxford, 2021). This report highlights the growing threat from hidden and systemic risks inherent in the emerging technology environment, which will require significant change to the international and security communities' response to cybersecurity.

*Gender Approaches to Cybersecurity: Design, Defense and Response* (United Nations Institute for Disarmament Research, 2021). The report explores how gender norms shape specific activities pertaining to cybersecurity design, defense, and response. The report proposes the incorporation of gender considerations throughout international cybersecurity policy and practice.

*The Global CSIRT Maturity Framework* (The Global Forum on Cyber Expertise, 2021). The Global Computer Security Incidence Response Team (CSIRT) Maturity Framework is intended to enhance global cyber incident management capacity.

*Global Good Practices: IoT Security* (Netherlands Organization for Applied Scientific Research, Singapore Cybersecurity Agency, 2019). IoT technologies can improve government operations, support better living, create new business opportunities, and support safer communities. But consumer privacy and safety can be undermined by the vulnerability of individual devices, connectivity, and back-end infrastructure; the wider economy faces an increasing threat of large-scale cyber attacks launched from insecure IoT devices. The hazards that come with the introduction of IoT must be managed to risk levels acceptable to society.

*Global Overview of Existing Cyber Capacity Assessment Tools* (GFCE, 2021). The report provides a comprehensive overview of the different cyber capacity assessment tools, their approaches, benefits, and outputs, and what to do and whom to contact if a country wishes to be assessed.

*Guide to Developing a National Cybersecurity Strategy* (ITU, 2021, 2nd Ed). This guide was developed by twenty partners from intergovernmental and international organizations, private sector, as well as academia and civil society.

*How to set up CSIRT and SOC: Good Practice Guide* (ENISA, NRD Cyber Security, 2020). This publication provides results-driven guidance for establishing a CSIRT or security operations center (SOC).

*International Strategy to Better Protect the Financial System Against Cyber Threats* (Carnegie Endowment, 2020). This report offers a vision for how the international community could better protect the financial system against cyber threats.

*Japan's Cybersecurity Policy: An Introduction (EU Cyber Direct, 2020).* Japan's cybersecurity policy and international engagements are developing rapidly. The Japanese government relies on public–private partnerships to tackle cybersecurity challenges. Japan's policy instruments are similar to those of the European Union (EU) and the US to address IoT security and supply chain security, such as certifications, minimum standards, and transparency requirements.

*National Capabilities Assessment Framework (ENISA, 2020).* This report presents the work performed by the European Union Agency for Cybersecurity (ENISA) to build a National Capabilities Assessment Framework.

The *Paris Call for Trust and Security in Cyberspace* of November 2018 is based on nine common principles to secure cyberspace.

*Playing with Lives: Cyberattacks on Healthcare are Attacks on People (Cyber Peace Institute, 2021).* This report focuses on the impacts of attacks on people and society. It consolidates scattered information, demonstrating the complexity, magnitude, and scope of the threat to health care, from ransomware through disinformation to COVID-19-related cyber espionage. The report analyzes the diversity in threat actors and their incentives, the difficult implementation of domestic and international norms and laws, and the under-resourcing of the sector despite a vibrant ecosystem of assistance initiatives. It shows how accountability is critical to any systemic solution: as incidents are under reported, attacks are seldom attributed and threat actors evade punishments.

*Putting Cyber Norms in Practice: Implementing the UN GGE 2015 recommendations through national strategies and policies (GFCE, 2021).* The implementation guide seeks to facilitate, inform, and promote collaborative and coordinated efforts to maintain and further develop an open, free, peaceful, and stable cyberspace through adequate national, regional, and global cybersecurity practices.

*Routledge Companion to Global Cyber-Security Strategy (Routledge, 2021).* This report provides the most comprehensive and up-to-date comparative overview of the cybersecurity strategies and doctrines of the major states and actors in Europe, North America, South America, Africa, and Asia. The volume offers an introduction to each nation's cybersecurity strategy and policy, along with a list of resources.

*Ransomware Task Force—Combating Ransomware: A Comprehensive Framework for Action (Institute for Security and Technology, 2021).* Ransomware is no longer just a financial crime; it is an urgent national security risk that threatens schools, hospitals, businesses, and governments across the globe. Over 60 experts from industry, government, law enforcement, civil society, and international organizations worked together to produce this comprehensive framework for a unified public-private anti-ransomware campaign.

*Safeguarding our Healthcare Systems: A Global Framework for Cybersecurity (Leading Health Systems Network, 2020).* This report looks at existing cybersecurity frameworks worldwide and examines why the health-care sector remains one of the worst adopters of cybersecurity frameworks. Members of the Leading Health Systems Network (LHSN) explored the current cybersecurity landscape and developed the Essentials of Cybersecurity in Healthcare Organizations (ECHO) framework.

*Smart and Safe: Risk Reduction in Tomorrow's Cities* (East-West Institute, 2019). This report has been developed to provide guidance and support to key decision-makers, including municipalities, governments, urban planners, businesses, and community leaders—to make tomorrow's smart cities secure and safe by managing technology effectively. The guide provides recommended actions across four major areas: cybersecurity, cyber resilience, privacy and data protection, and collaboration and coordination in governance.

*Software Supply Chain Attacks* (CISA & NIST, 2021). The report provides an overview of software supply chain risks and recommendations on how software customers and vendors can use the NIST Cyber Supply Chain Risk Management Framework and the Secure Software Development Framework to identify, assess, and mitigate software supply chain risks. This resource provides in-depth recommendations for software customers and vendors as well as key steps for prevention, mitigation, and resilience of software supply chain attacks.

*Towards Identifying Critical National Infrastructures in the National Cybersecurity Strategy Process* (GFCE, 2022). This white paper builds upon existing Critical National Infrastructure (CNI/CII) work and proposes some practical considerations and measures for how countries can develop approaches for identifying CNI/CII as part of their NCS development and implementation processes.

*X-Force Threat Intelligence Index* (IBM, 2022). The annual IBM X-Force® Threat Intelligence Index sheds light on recent trends and offers recommendations to help bolster security strategies for the future.

# APPENDIX 4
# OPTIONS FOR GLOBAL ACTION ON CYBERCRIME

For governments attempting to prevent cybersecurity failures, patchwork enforcement mechanisms across jurisdictions continue to hamper efforts to control cybercrime. Geopolitical tensions hinder potential cross-border collaboration. Already suffering from a loss of public trust in the wake of the COVID-19 crisis, governments may face further societal anger if they are unable to both keep up with the shifting threat landscape and responsibly manage these challenges.

The Ransomware Taskforce recently proposed a comprehensive framework of 48 actions that government and industry can adopt at a global level to significantly disrupt the ransomware business model and mitigate the impact of these attacks in the immediate and long terms. Initiatives also need to focus on emerging technologies such as artificial intelligence and quantum, as well as the modes of digital exchange they make possible.

While multistakeholder international dialogues can help strengthen links between actors operating in the digital security space, striking the right balance between national security needs and the needs of enterprises remains elusive. For instance, the US Vulnerabilities Equities Process looks heavily weighted in favor of government interests at present, while keeping known vulnerabilities concealed from industry. Agreements like the Siemens Charter, which are meant to regulate the conduct of industry, remain very limited in scope with few clear principles behind them. They also require far more signatories from industry and more obvious sanctions to be effective. Similar doubts relate to more international facing frameworks, like the Paris Call. Absent as signatories are key cyber superpowers like the US, the Russian Federation, the People's Republic of China (PRC), Iran, Israel, and Democratic People's Republic of Korea.

Fully global agreements are essential, yet approaches are at best limited, at worst lacking in several important dimensions. For instance, the proposal by the Russian Federation and the PRC for a cybercrime treaty would permit internet blackouts and the criminalization of free speech.

Table A4 summarizes the kinds of agreements available for regulating conduct *prior* to the outbreak of more advanced forms of cyber conflict, relying on varying degrees of securing better cooperation and more active conflict prevention. The Tallin Manual offers a very detailed discussion for the kinds of agreements available for regulating conduct should more advanced forms of cyber conflict occur.

## Table A4: Current and Proposed Frameworks for Cyber Treaties

| Level and Key Actors | Examples of Current/Proposed Agreement | Scope | Content |
|---|---|---|---|
| **Standards of Conduct/Consensus prior to Advanced Cyberconflict** | | | |
| **Domestic** | US Vulnerability Equities Process (VEP) | Transparency and cooperation between state and government agencies in reducing cyber threats | A process used by the US government to determine on a case-by-case basis how it should treat zero-day computer security vulnerabilities: whether to disclose them to the public to help improve computer security or keep them secret for offensive use against adversaries. VEP requires the US government to disclose cyber vulnerabilities to leading tech companies, while permitting intelligence agencies to retain information about zero-days.[a] |
| **Industry** | Cybersecurity Tech Accord | Digital tech industry cooperation to reduce cyber threats | Aims to foster cooperation among digital tech companies to enhance the "security, stability and resilience of cyberspace" especially against nation-state threats. Over 150 signatories, including HP, Meta, Dell, BT, Microsoft, Hitachi, Panasonic, Cisco, Nokia, RSA, and Orange. |
| **Industry** | Siemens Charter of Trust | Creation of industry standards for online safety against cyber threats. | Aims to "set minimum general standards for cybersecurity" centered on three goals for safer networks for individuals, companies, and infrastructures: data protection; damage limitation; reliable foundations for trust. |
| **Industry and States** | Paris Call for Trust and Security in cyberspace (2018) | Cooperation between industry and states against cyber threats | Calls to enhance international safety against malicious online activity, preventing interference in electoral processes, tightening up of online mercenary activities and offensive action by non-state actors, and cooperating to enhance international standards. Signed by 81 countries, 700+ companies, and 390 CSOs. |
| **States and/or International** | Budapest Convention on Cybercrime | Enhancing consensus and cooperation in fighting cybercrime | Ratified by 67 states, the convention aims to harmonize domestic law and policy toward online offending among signatory states. The convention is oriented more toward crime than norms of conduct and/or Russian Federation, India, and the People's Republic of China have refused to sign up because of strategic concerns. |
| **States and/or International** | (Proposed) UN Cybercrime Treaty | Enhancing consensus and cooperation in fighting cybercrime | Proposed by the Russian Federation, the People's Republic of China, and others in early 2020. An ad hoc committee of experts, which represent all regions, is to elaborate details on a 'comprehensive international convention on countering the use of information and communications technologies for criminal purpose.[b] |
| **States and/or International** | UN GGE Process | Creating consensus about norms of cyber conduct | UN Group of Governmental Experts, which aims to develop greater consensus between all member states on acceptable conducts and norms across digital networks.[c] |

*continued on next page*

| Level and Key Actors | Examples of Current/Proposed Agreement | Scope | Content |
|---|---|---|---|
| **Standards of Conduct/Consensus during Advanced Cyberconflict** | | | |
| **State and/or International** | Tallin Manual 2.0 | Creating standards for nation-state conduct during cyber operations | Composed by international legal experts, the Tallin Manual 2.0 provides an objective restatement of international law as applied to cyber operations that states encounter and that fall below the thresholds of the use of force or armed conflict. The manual is policy and politics neutral; considers how existing principles surrounding "digital sovereignty, jurisdiction, human rights, due diligence, and the prohibition of intervention" might apply to each given scenario.d A 5-year update process is underway, with Tallin Manual 3.0 expected by 2026. |

[a] For an in-depth analysis of benefits and shortcomings of the Vulnerability Equities Process see L. Polley. 2022. To Disclose, or Not to Disclose, That Is the Question: A Methods-Based Approach for Examining & Improving the US Government's Vulnerabilities Equities Process. Santa Monica, California: RAND Corporation.

[b] Following a first round of consultations in 2022, observers noted a pronounced lack of consensus among UN member states about what constitutes a "cybercrime" and how expansive the treaty will be. K. Rodriguez and M. Baghdasaryan. 2022. UN Committee To Begin Negotiating New Cybercrime Treaty Amid Disagreement Among States Over Its Scope. *Electronic Frontier Foundation*. 15 February.

[c] For background on the outcome of the UN Group of Governmental Experts' Process, see ICT4Peace Foundation. ICT4Peace Commentary on Final UN GGE 2021 Report.

[d] For insights on the Tallin Manual 2.0, see E. T. Jensen. 2017. The Tallinn Manual 2.0: Highlights and Insights. 48 Georgetown Journal of International Law 735 (2017), BYU Law Research Paper No. 17–10. *SSRN Scholarly Paper*. Rochester, New York. Social Science Research Network.

Source: M. McGuire. 2021. Nation States, Cyberconflict, and the Web of Profit. *HP Wolf Security* (blog). 8 April.

# REFERENCES

Abdelsalam, M. et. al. 2020. Security and Privacy in Smart Farming: Challenges and Opportunities. *IEEE Access* 8: 34564–84.

Achten, N. et. al. 2020. *Principled Artificial Intelligence: Mapping Consensus in Ethical and Rights-Based Approaches to Principles for AI*. Boston: Harvard University Berkman Klein Center for Internet and Society.

Ada Lovelace Institute. Algorithmic Impact Assessment in Healthcare.

Adam, T. 2019. Digital Neocolonialism and Massive Open Online Courses (MOOCs): Colonial Pasts and Neoliberal Futures. *Learning, Media and Technology*. 44 (3). pp. 365–80.

Adomaitis, L. and A. Grinbaum. 2023. Dual Use Concerns of Generative AI and Large Language Models. *arXiv*. 2305.7882.

Ajijola, A. and N.D.F. Allen. 2022. African Lessons in Cyber Strategy. *Africa Center for Strategic Studies*. 8 March.

Alicke, K. et. al. 2020. Risk, Resilience, and Rebalancing in Global Value Chains. McKinsey. 6 August.

Allianz. Allianz Risk Barometer.

Amazon Web Services (AWS). 2021. Why Moving to the Cloud Should Be Part of Your Sustainability Strategy. *AWS Public Sector Blog*. 25 October.

American Psychological Association. Resilience.

Ariel, S. et. al. 2022. Global Data Governance Mapping Project.

Arthur D. Little. Technology Foresight: Anticipating Future Impact.

Arya, S. and S. Kumar. 2020. E-Waste in India at a Glance: Current Trends, Regulations, Challenges and Management Strategies. *Journal of Cleaner Production*.

Ashtari, H. 2022. What Is Network Behavior Analysis? Definition, Importance, and Best Practices. *Spiceworks* (formerly Toolbox). 15 February.

Asian Development Bank. 2020. *Futures Thinking in Asia and the Pacific: Why Foresight Matters for Policy Makers*.

Asian Development Bank. 2022. *Reimagining the Future of Transport across Asia and the Pacific*.

Asian Development Bank and AWS Institute. 2023. *Data Management Policies and Practices in Government*.

Association for Progressive Communications. 2021. Community Networks from Asia, Africa and Latin America Gather Online to Share Experiences, Learnings and Resources. News release. 5 August.

AuditBoard. 2022. Digital Risk Maturity Report 2022: Turning Digital Risk Into Your Competitive Advantage.

Axiata. Axiata Sustainability–Advancing To Zero.

Azure. 2020. 10 Recommendations for Cloud Privacy and Security with Ponemon Research. *Microsoft Azure* (blog). 29 January.

Baghdasaryan, K. and M. Rodriguez. 2022. UN Committee To Begin Negotiating New Cybercrime Treaty Amid Disagreement Among States Over Its Scope. *Electronic Frontier Foundation*. 15 February.

Baker, M.G. 2020. Successful Elimination of Covid-19 Transmission in New Zealand. *The New England Journal of Medicine*. 383 e56.

Baker McKenzie. 2021/2022 Digital Transformation & Cloud Survey: A Wave of Change.

Baldé, C.P. et. al. The Global E-Waste Monitor 2020.

Banaji, S. and R. Bhat. 2020. India: The Dissemination of Misinformation on WhatsApp Is Driving Vigilante Violence against Minorities. *Minority Rights Group*. 29 September.

Barrett, B. 2020. Marriott Got Hacked. Yes, Again. *Wired*. 31 March.

Barry, E. 2023. Social Media Use Is Linked to Brain Changes in Teens, Research Finds. *The New York Times*. 3 January.

Barry, R. and D. Volz. 2019. Ghosts in the Clouds: Inside China's Major Corporate Hack. *The Wall Street Journal*. 30 December.

Bartol, N. et. al. 2011. Key Practices in Cyber Supply Chain Risk Management: Observations from Industry. *National Institute of Standards and Technology*. 11 February.

Bean, R. et. al. 2021. Why Do Chief Data Officers Have Such Short Tenures? *Harvard Business Review*. 18 August.

Beato, F. et. al. 2021. Cyber Resilience in the Oil and Gas Industry: Playbook for Boards and Corporate Officers. World Economic Forum, Siemens Energy, and Saudi Aramco. 17 May.

Bennett, S. 2022. Cloud File Security Statistics 2022. *Webinar Care*. 3 October.

Berg, L. 2021. RTF Report: Combating Ransomware. *Institute for Security and Technology*.

Bergman, R. and M. Mazzetti. 2022. The Battle for the World's Most Powerful Cyberweapon. *The New York Times*. 28 January.

Bhakta, A. and J. Wheeler. 2022. The Business Resilience Gap: A Tipping Point. *AuditBoard* (blog). 1 February.

Bixal. Building Cybersecurity Capacity of Civil Society Organizations in Colombia to Improve Digital Health and Protect against Cyber Threats.

Block, C. 2022. Council Post: 12 Reasons Your Digital Transformation Will Fail. *Forbes*. 16 March.

Bordage, F. 2019. Study-The Environmental Footprint of the Digital World.

Bozikovic, A. 2022. An Inside Look at How Alphabet's Sidewalk Labs Failed in Toronto. *Architectural Record*. 22 March.

Brende, B. and B. Sternfels. 2022. Resilience for Sustainable, Inclusive Growth. 7 June.

Brooks, S. et. al. 2017. An Introduction to Privacy Engineering and Risk Management in Federal Systems. *National Institute of Standards and Technology*.

Brown, D. 2021. Bitcoin Miners Break New Ground in Texas, a State Hailed as the New Cryptocurrency Capital. *Washington Post*. 8 July.

Brown, L. 2021. The New Threat Economy: A Guide to Cybercrime's Transformation – and How to Respond. *Security Magazine*. 15 June.

Brundyn, A. et. al. 2022. Using Federated Learning to Bridge Data Silos in Financial Services. *NVIDIA Technical Blog*. 16 August.

Bundesministerium für Bildung und Forschung. 2021. Insight-BMBF.

Burman, A. 2022. The Withdrawal of the Proposed Data Protection Law Is a Pragmatic Move. *Carnegie India.*

Business & Human Rights Resource Centre. 2011. UN Guiding Principles.

Business and Human Rights Resource Centre. 2021. German Parliament Passes Mandatory Human Rights Due Diligence Law.

BW0 Italy. 2021. B20 Final Communique: Policy Recommendations to the G20.

Caldwell, J.H. 2021. A Moving Target: Refocusing Risk and Resiliency amidst Continued Uncertainty. *Deloitte Insights.* 1 February.

Calleja-Sanz, G. et. al. 2020. Technology Forecasting: Recent Trends and New Methods. In *Research Methodology in Management and Industrial Engineering,* edited by Carolina Machado and J. Paulo Davim. Springer International Publishing.

Cannon, M. et al. 2020. How Mental Health Care Should Change as a Consequence of the COVID-19 Pandemic. *The Lancet. Psychiatry.* 7 (9): 813–24.

Carnegie Endowment for International Peace. Timeline of Cyber Incidents Involving Financial Institutions.

Carter, N. 2021. How Much Energy Does Bitcoin Actually Consume? *Harvard Business Review.* 5 May.

Cavoukian, A. 2011. Privacy by Design: The 7 Foundational Principles.

CEE Bankwatch Network. 2021. Raw Deal.

Center for AI Safety. 2023. Statement on AI Risk.

Center for Global Development. Governing Data for Development: Trends, Challenges, and Opportunities.

Center for Strategic and International Studies. Significant Cyber Incidents.

Ceres, P. 2022. The US May Soon Learn What a "Kid-Friendly" Internet Looks Like. *Wired.* 1 September.

Chakraborty, S. et. al. 2020. Flipping the Odds of Digital Transformation Success. *Boston Consulting Group Global.* 29 October.

Chaparro-Banegas, N. et. al. 2021. Digital Transformation: An Overview of the Current State of the Art of Research. *SAGE Open.* 11 (3): 21582440211047576.

Chen, Q. et. al. 2021. Seven Major Changes in China's Finalized Personal Information Protection Law. *Stanford University DigiChina Project.* 15 September.

Cheong, N. 2022. Design a Federated Learning System in Seven Steps. *Integrate.ai.* 16 June.

Chew, B. et. al. 2020. How Governments Can Navigate a Disrupted World. *Deloitte Insights.* 24 July.

CIO Summits. Managing Digital Risk: 8 Types of Digital Risk and How to Manage Them.

Cisco. 2022. Cisco 2022 Data Privacy Benchmark Study.

Cities Coalition for Digital Rights. Declaration.

Cities Coalition for Digital Rights. Digital Rights Governance Framework.

Clarke, S. and A. Annabelle. 2019. The Ethics of Artificial Intelligence: Laws from around the World. *MinterEllison.* 3 June.

Climate Care. The Carbon Footprint of the Internet.

Clinton, K. and K. Lewis. Panel: The Key to Third-Party Risk Management in APAC-Aligned Assurance. *Gartner* (webinar). On demand.

Cotruvo, J. et. al. 2021. Bridging Hydrometallurgy and Biochemistry: A Protein-Based Process for Recovery and Separation of Rare Earth Elements. *ACS Central Science.* 7 (11).: 1798–1808.

Council of Europe. 2021. Children's Data Protection in an Education Setting-Guidelines.

Council of Europe. 2018. Guidelines to respect, protect and fulfil the rights of the child in the digital environment.

Council of Europe. Global Action on Cybercrime Extended (GLACY)+.

Council of Europe. The 24/7 Network Established under the Budapest Convention on Cybercrime.

Council for Registered Ethical Security Testers (CREST). Cyber Security Incident Response Maturity Assessment.

CREST. Who Are CREST?

CREST. CREST Member Companies.

CrowdStrike. What Is Cloud Data Security? Risks & Best Practices.

Cruz, L. et. al. 2021. AI Lifecycle Models Need to Be Revised. *Empirical Software Engineering.* 26 (5): 95.

Culp, S. 2022. Confronting the Risks of Innovation and Technology. *Forbes.* 27 September.

Culp, S. 2021. In a World of Risk, Pace Comes from Preparation. *Accenture.* 13 July.

Crypto Climate. Crypto Climate Accord.

Cyber Peace Institute. 2022. Cyber Attacks in Times of Conflict.

Cyber Resilience for Development. We are Cyber4d.

Czerniewicz, L. 2018. Unbundling and Rebundling Higher Education in an Age of Inequality. *Educause Review.* 29 October.

Dabbs, R. 2019. How Can Your Digital Strategy Help Improve EHS Outcomes? *Ernst & Young.* 12 August.

Dahlman, C. et. al. 2021. The Converging Technology Revolution and Human Capital: Potential and Implications for South Asia. World Bank.

Dan, X. et. al. 2023. How Will China's Generative AI Regulations Shape the Future? A DigiChina Forum. *Stanford University DigiChina Project.* 19 April.

Data Privacy Manager. 2021. Data Privacy vs. Data Security (Definitions and Comparisons).

Dawande, M. et. al. 2022. This Is the Key to Designing Sustainable Data Cooperatives. *World Economic Forum.* 1 February.

Dawson, R. 2022. Government Foresight Programs.

Deasy, D. et. al. 2022. Integrated National Data Ecosystems: The next Stage of Digital Transformation. *World Bank Blogs.* 6 September.

Dekker, T. 2021. The Board Imperative: How Can Data and Tech Turn Risk into Confidence? *Ernst & Young.* 14 July.

Dekker, T. et. al .2021. The Board Imperative: Is Now the Time to Reframe Risk as Opportunity? *Ernst & Young.* 14 July.

Deloitte. Perspective: Digital Government Transformation.

Deloitte Insights. Government Trends 2023.

Dener, C. et. al. 2021. *GovTech Maturity Index: The State of Public Sector Digital Transformation.* World Bank.

Deutsche Telekom. 2020. Climate Strategy: 2020 Corporate Responsibility Report.

Development Data Partnership. About the Development Data Partnership.

Digital Europe. 2022. Data Centres: Powering the Green and Digital Transition in Europe. YouTube webinar. 1:26.56. 12 February.

Digital Impact Alliance. 2022. Comparative Analysis of Digital Transformation Funding and Financing Models. 1 April.

Dilmegani, C. 2021. What Is Homomorphic Encryption? Benefits & Challenges [2022]. *AI Multiple.* 19 August.

Dodson, D. et al. 2022. Secure Software Development Framework Version 1.1: Recommendations for Mitigating the Risk of Software Vulnerabilities. *National Institute of Standards and Technology Special Publication.* 800–218.

Douglas, E. et. al. 2021. The Outcomes of Sexting for Children and Adolescents: A Systematic Review of the Literature. *Journal of Adolescence* 92 (1): 86–113.

Ducker, J. 2022. Investigating the Vulnerability of the Food Industry to Cyberattacks. *AZO Life Sciences*. 4 March.

Duong, V. 2021. Vendor Selection Criteria: 7 Key Features To Keep In Mind. *Technology Insider*. 18 May. DSpark.

Edwards, L. 2022. Expert Opinion: Regulating AI in Europe. *Ada Lovelace Institute*. 31 March 31.

Elastic. Log Monitoring with Elastic Observability.

Engler, A. 2022. The EU AI Act Will Have Global Impact, but a Limited Brussels Effect. *Brookings*. 8 June.

A. Feigenbaum et. al. 2022. Data Governance, Asian Alternatives: How India and Korea Are Creating New Models and Policies. *Carnegie Endowment for International Peace*. 31 August.

Erin. 2022. Google Analytics Privacy Issues: Is It Really That Bad? *Matomo* (blog). 2 June.

Espinoza, J. 2022. EU to Impose Tough Rules on 'Internet of Things' Product Makers. *Financial Times*. 8 September.

Ethereum. Proof-of-Stake (PoS).

EU CyberNet. EU CyberNet-the bridge to cybersecurity expertise in the European Union.

EU Policy Lab. 2023. Technology Foresight: Anticipating the Innovations of Tomorrow. 26 May.

EUR-Lex. Proposal for a Regulation of the European Parliament and of the Council Laying Down Harmonised Rules on Artificial Intelligence (Artificial Intelligence Act) and Amending Certain Union Legislative Acts.

European Bank for Reconstruction and Development. The EBRD's Digital Approach.

European Commission. 2020 Strategic Foresight Report.

European Commission. A European Approach to Artificial Intelligence.

European Commission. Declaration on European Digital Rights and Principles.

European Commission. Lessons from US NIST for EU AI Act-Future of Life Institute.

European Commission. The EU Cybersecurity Act.

European Cyber Security Organisation. European Cyber Security Investment Platform.

European Union. General Data Protection Regulation (GDPR) Compliance Guidelines.

Evans, R. et. al. 2021. Cyberbullying Among Adolescents and Children: A Comprehensive Review of the Global Situation, Risk Factors, and Preventive Measures. *Frontiers in Public Health* 9.

Federation of European Risk Management Associations and McKinsey. 2021. The Role of Risk Management in Corporate Resilience. *Federation of European Risk Management Associations–FERMA*. 10 June.

Felländer, A. et. al. 2022. Achieving a Data-Driven Risk Assessment Methodology for Ethical AI. *Digital Society*. 1 (2): pp. 1–27.

Fergnani, A. 2022. Corporate Foresight: A New Frontier for Strategy and Management. *Academy of Management Perspectives*. 36 (2): 820–44.

Flynn, S. 2022. Machine Learning Life Cycle: 6 Stages Explained. *CIO Insight*. 22 July.

Foitzik, P. 2020. How to Manage Privacy Risk under Both the GDPR and CCPA. *International Association of Privacy Professionals*. 25 February.

Franceschi-Bicchierai, L. 2022. Facebook Doesn't Know What It Does With Your Data, Or Where It Goes: Leaked Document. *Vice* (blog). 26 April.

Fraunhofer Institute for Reliability and Microintegration IZM, Berlin. Proceedings of the Sixth Electronics Goes Green Conference.

Freed, B. 2021. 44% of Education Institutions Targeted by Ransomware in 2020, Survey Finds. *EdScoop*. 23 July.

Freedom House. Freedom on the Net 2021: The Global Drive to Control Big Tech.

Freedom House. Freedom on the Net 2022: Countering an Authoritarian Overhaul of the Internet.

Future Today Institute. What We See.

Gartner. 7 Top Trends in Cybersecurity for 2022.

Gartner. Gartner Information Technology Glossary: Definition of Integrated Risk Management.

Gartner. Maximize Third-Party Risk Management With Aligned Assurance. *Gartner* (webinar). On demand.

Gartner. TPRM Governance and Technology Investments.

Gartner. Where and How to Target your Digital Business Transformation.

Geneva Environment Network. The Growing Environmental Risks of E-Waste.

Global Cyber Security Capacity Centre. Cybersecurity Capacity Maturity Model for Nations 2021 Edition.

Global Data and Marketing Alliance. Global data privacy: What the consumer really thinks.

Global Forum on Cyber Expertise. News.

Global Lancers. Top 7 Reasons Why Digital Transformations Fail.

Global Legal Insights. AI, Machine Learning & Big Data Laws and Regulations: China.

Global Privacy Assembly. Adopted Resolutions.

Global Privacy Assembly. List of Accredited Members.

Gorley, A. 2022. What Is ESG and Why It's Important for Risk Management. *Sustainalytics*. 2 March.

Government of Australia, Cyber and Infrastructure Security Centre. Critical Infrastructure Resilience Strategy.

Government of Australia, Department of Foreign Affairs and Trade. 2021 International Cyber and Critical Technology Engagement Strategy.

Government of Australia, Office of the Australian Information Commissioner. Guide to Undertaking Privacy Impact Assessments.

Government of Canada, Office of the Privacy Commissioner. 2019. PIPEDA Compliance Help. 11 July.

Government of Israel, National Cyber Directorate. Israel International Cyber Strategy.

Government of Malaysia. National Cyber Security Agency.

Government of Singapore, Personal Data Protection Commission. Guide to Data Protection Impact Assessments.

Government of Singapore, Personal Data Protection Commission. ASEAN Data Management Framework and Model Contractual Clauses on Cross Border Data Flows. January 2021.

Government of the United States, Cybersecurity and Infrastructure Security Agency. Critical Infrastructure Sectors.

Government of the United States, Department of Energy. Cybersecurity Capability Maturity Model.

Government of the United States, Government Accountability Office. 2022. Critical Infrastructure Protection: CISA Should Improve Priority Setting, Stakeholder Involvement, and Threat Information Sharing. Press release. 1 March.

Government of the United States, National Institute of Standards and Technology (NIST). AI Risk Management Framework.

Government of the United States, National Institute of Standards and Technology. Cybersecurity Framework 2.0.

Government of the United States, National Institute of Standards and Technology. Cybersecurity Framework Documents.

Government of the United States, National Institute of Standards and Technology. NIST Special Publication 800–30.

Government of the United States, National Institute of Standards and Technology. Privacy Framework.

Government of the United States, National Institute of Standards and Technology. Software Security in Supply Chains: Software Bill of Materials.

Government of the United States, Office of the Director of National Intelligence. 2021 Annual Threat Assessment of the U.S. Intelligence Community.

Government of the United States, Office of the Director of National Intelligence. 2022 Annual Threat Assessment of the U.S. Intelligence Community.

GRC World Forums. 2021. Ransomware Demands Soar by 518% in 2021. 13 August.

Greenleaf, G. 2022. Now 157 Countries: Twelve Data Privacy Laws in 2021/22. *176 Privacy Laws & Business International Report*. 1, 3–8.

Greis, J. et. al. 2022. Software Bill of Materials: Managing Software Cybersecurity Risks. McKinsey. 19 September.

Gross, J. 2023. How Finland Is Teaching a Generation to Spot Misinformation. *The New York Times*. 10 January.

GSMA. 2021. ClimateTech. *Mobile for Development* (blog).

GSMA. 2021. The GSMA Sustainability Assessment Framework 2021. *#BetterFuture* (blog).

Guth-Orlowski, S. 2021. The Digital Product Passport and Its Technical Implementation. *Medium* (blog). 2 December.

Han, H. 2022. How Dare They Peep into My Private Life? *Human Rights Watch*. 25 May 25.

Hankins, J. 2015. What Does "Responsible Innovation" Mean? *IEEE Spectrum*. 24 June. J. Wheeler. 2020.

Hanna, N.K. 2020. Assessing the Digital Economy: Aims, Frameworks, Pilots, Results, and Lessons. *Journal of Innovation and Entrepreneurship*. 9 (1): 16.

Heeks, R. 2016. RABIT: A New Toolkit for Measuring Resilience. *Nexus for ICTs, Climate Change and Development*. 14 November.

Heikkilä, M. 2022. A Quick Guide to the Most Important AI Law You've Never Heard of. *MIT Technology Review*. 13 May.

Heo, H. et. al. 2020. COVID-19 in South Korea: Lessons for Developing Countries. *World Development*. 135: 105057.

Hebda, D. 2021. Legal and Compliance Risk Hot Spots for 2022. Gartner Webinar. November.

Hill, M. 2022. The Apache Log4j Vulnerabilities: A Timeline. *CSO Online*. 7 January.

Himmelreich, H. and I. Oshri. 2020. Your Digital Transformation Needs a Smart Vendor Strategy. *Boston Consulting Group Global*. 30 October.

Holloway, K. et. al. 2021. Digital Identity, Biometrics and Inclusion in Humanitarian Responses to Refugee Crises. *ODI*. 6 October.

Hotel Tech Report. 2020. Marriott Data Breach FAQ: What Really Happened? 9 December.

Howell, C. 2022. AI Regulation: Where Do China, the EU, and the US Stand Today? *The National Law Review*. 2 August.

Huang, J. et. al. 2021. Bitcoin Uses More Electricity Than Many Countries. How Is That Possible? *The New York Times*. 3 September.

Huls, A. 2021. To Improve Higher Ed Data Security, Address These Risks in Research Projects. *EdTech Magazine.* 10 May.

IBM Institute for Business Value. 2011. Digital Transformation.

ICT4Peace. 2021. ICT4Peace Commentary on Final UN GGE 2021 Report. 12 June.

IFRS Foundation. International Sustainability Standards Board.

Institute for Health Metrics and Evaluation. 2022. Lack of Trust Has Helped Fuel the COVID-19 Pandemic, New Study Shows. News release. 31 January.

Institute for Security & Technology. 2022. Combating Ransomware - A Comprehensive Framework for Action: Key Recommendations from the Ransomware Task Force.

Integrated Reporting. Integrated Reporting Framework.

Inter-American Development Bank. 2021. A Cybersecurity Guide for Smart Cities.

International Association of Privacy Professionals. 2022. Measuring Privacy Programs: The Role of Metrics. 14 March.

International Centre for Missing & Exploited Children. ITU Child Online Protection Guidelines 2020.

International Committee of the Red Cross. 2022. Cyber-Attack on ICRC: What We Know. 21 January.

International Committee of the Red Cross. 2022. Sophisticated Cyber-Attack Targets Red Cross Red Crescent Data on 500,000 People. News release. 19 January.

International Data Corporation. 2021. New IDC Spending Guide Shows Continued Growth for Digital Transformation as Organizations Focus on Strategic Priorities.

International Data Corporation. 2022. Worldwide Digital Transformation Investments.

International Institute for Applied Systems Analysis. The Digital Revolution and Sustainable Development: Opportunities and Challenges: Report Prepared by The World in 2050 Initiative.

International Panel on the Information Environment. Scientists.

International Telecommunication Union. Benchmark of Fifth-Generation Collaborative Digital Regulation. 2022, 93.

International Telecommunication Union. Connecting Humanity.

Investor Alliance for Human Rights. Investor Statement on Corporate Accountability for Digital Rights. *IoT Analytics.* 18 May.

ISMS.online. ISO 27701, The Privacy Information Management Standard.

ISO. ISO 31000:2009(En), Risk Management — Principles and Guidelines.

ISO/IEC. ISO/IEC 27005 Risk Management.

IT for Change. Data Governance Network.

ITU Hub. G5 Regulation: The Digital Transformation Fast Lane.

Iyengar, P. Roadmap to Renewal: Insights from the 2022 Board of Directors Survey. *Gartner* (webinar). On demand.

Iyengar, R. 2023. The Global Race to Regulate AI. *Foreign Policy* (blog). 5 May.

Jensen, E.T. 2017. The Tallinn Manual 2.0: Highlights and Insights. *SSRN Scholarly Paper.* Social Science Research Network.

Jerath, K. 2022. Mobile Advertising and the Impact of Apple's App Tracking Transparency Policy.

Johnson, K. 2022. Europe Prepares to Rewrite the Rules of the Internet. *Wired.* 28 October.

Johnston, A. 2022. The Seven Habits of Effective Privacy Impact Assessments. *Salinger Privacy.* 2 August. 2022.

Kajáti, E. et. al. 2018. A Technology Selection Framework for Manufacturing Companies in the Context of Industry 4.0. In *2018 World Symposium on Digital Intelligence for Systems and Machines (DISA)*, 267–76.

Kaplan, J. et. al. 2022. Creating a technology risk and cyber risk appetite framework. McKinsey.

Kaplan, J. et. al. 2022. From Risk Management to Strategic Resilience. March.

Kaplan, J. et. al. 2022. McKinsey on Risk. April.

Kaplan, J. et. al. 2019. Perspectives on Transformation.

Kitani. 2019. The $900 Billion Reason GE, Ford and P&G Failed at Digital Transformation. *CNBC*. 30 October.

Kleinman, L. 2020. Council Post: The Rise of Third-Party Digital Risk. *Forbes*. 14 July.

Kohler, K. *Estonia's National Cybersecurity and Cyberdefense Posture: Policy and Organizations*. ETH Zürich Center for Security Studies.

Koomey, J. et. al. 2020. Recalibrating Global Data Center Energy-Use Estimates. *Science*. 367 (6481): 984–86.

Korte, M. 2020. The Impact of the Digital Revolution on Human Brain and Behavior: Where Do We Stand? *Dialogues in Clinical Neuroscience*. 22 (2): 101–11.

Kost, E. 2022. Meeting the 3rd-Party Risk Requirements of The NY SHIELD Act. *UpGuard*. 22 August.

KPMG. 2022. A Triple Threat across the Americas: 2022 KPMG Fraud Outlook.

KPMG. 2022. Third-Party Risk Management Outlook 2022.

Krebs, B. 2014. Target Hackers Broke in Via HVAC Company. *Krebs on Security*. 14 February.

LEGO Group. Responsibility Report 2018.

Lichtenstein, S. 2022. When a Country Is Held Hostage: What to Make of Costa Rica's Ransomware Attack. *Risk Assistance Network Exchange-Stratfor*. 26 May.

Loehr, T. 2021. 8 Best Practices for Securing Infrastructure as Code. *Cycode* (blog). 6 October.

Loohuis, K. 2021. The Netherlands Still Lacks Digital Resilience, Says Report. *ComputerWeekly*. 11 August.

McCain, A. 2022. How Fast Is Technology Advancing? [2022]: Growing, Evolving, and Accelerating at Exponential Rates. *Zippia Research*. April.

McGuire, M. 2021. Nation States, Cyberconflict and the Web of Profit | HP Threat Research. *HP Wolf Security* (blog). 8 April.

McKinsey. 2022. Cybersecurity Trends: Looking over the Horizon. 10 March.

McKnight, C. 2022. Step by step is the best approach to a successful technology selection. *Digital Clrity Group* (blog).

Metricstream. The Comprehensive Guide To Integrated Risk Management.

Microsoft Ignite. Responsible Innovation Toolkit-Azure Application Architecture Guide.

Mordor Intelligence. Digital Transformation Market Size, Share (2022–2027): Industry Trends.

MuleSoft. 2021. Study Reveals Integration Challenges Threaten Digital Transformation, With Organizations Spending on Average $3.5 Million on Custom Integration Labor Costs.

Mutung'u, G. and F. Ogonjo. 2021. *Simplified Data Protection Impact Assessment for Small Organisations*. Nairobi: Strathmore University Center for Intellectual Property and Technology Law.

Myra Security. What Is IT Compliance?

National Cyber Security Index. NCSI Ranking.

National Institute of Standards and Technology. Building the NIST AI Risk Management Framework: Workshop #2.

Nauck, F. et. al. 2021. The Resilience Imperative: Succeeding in Uncertain Times. McKinsey. 17 May.

NAVEX. 2022. Definitive Risk and Compliance Benchmark Report.

Nederland Digitaal. The Dutch Digitalisation Strategy 2021.

Nelson, J. 2023. Take AI Warnings Seriously, Says UN Secretary-General. *Decrypt*. 14 June.

Newman, L.H. 2018. The Leaked NSA Spy Tool that Hacked the World. *Wired*. 7 March.

Newman, L.H. 2021. A Year After the SolarWinds Hack, Supply Chain Threats Still Loom. *Wired*. 8 December. (accessed 17 September 2022).

Newton, C. 2023. Microsoft Lays off Team That Taught Employees How to Make AI Tools Responsibly. *The Verge*. 14 March.

NMIP Algorithmic Impact Assessment (AIA) Template.

Nobel Prize. The Nobel Prize Summit: Truth, Trust and Hope.

Office of Technology Assessment at the German Bundestag. About Us.

Oladimeji, S. 2022. SolarWinds Hack Explained: Everything You Need to Know. *TechTarget*. 29 June.

Olmstead, L. 2021. 9 Critical Digital Transformation Challenges to Overcome (2022). *Whatfix (blog)*. 2 December.

Open Data Institute. Mapping Data Ecosystems: Methodology.

Open Data Institute. 2021. The Economic Impact of Trust in Data Ecosystems. 9 March.

Organisation for Economic Co-operation and Development. Data Governance.

Organisation for Economic Co-operation and Development. Encouraging Vulnerability Treatment: Overview for Policy Makers.

Organisation for Economic Co-operation and Development. 2015. *Digital Security Risk Management for Economic and Social Prosperity*. Paris: OECD.

Organisation for Economic Co-operation and Development. 2022. *OECD Policy Framework on Digital Security*. Paris: OECD.

Organisation for Economic Co-operation and Development. Privacy.

Organisation for Economic Co-operation and Development. 2019. Policy Toolkit on Governance of Critical Infrastructure Resilience. In *Good Governance for Critical Infrastructure Resilience*. Paris: OECD.

Organisation for Economic Co-operation and Development. 2020. Protecting Children Online: An Overview of Recent Developments in Legal Frameworks and Policies. Paris: OECD.

Organisation for Economic Co-operation and Development. Leaving no one behind in a digital world: the United Kingdom's Digital Access Programme.

Organisation for Economic Co-operation and Development. Foresight and Anticipatory Governance in Practice.

Organization for Security and Co-operation in Europe. OSCE Cyber Security Awareness Month. October 2020.

Osborne, C. 2021. Updated Kaseya Ransomware Attack FAQ: What We Know Now. *ZDNET*. 23 July.

O'Shaughnessy, M. 2023. What a Chinese Regulation Proposal Reveals About AI and Democratic Values. *Carnegie Endowment for International Peace*, 16 May.

Pacific Islands Forum. BOE Declaration Action Plan.

Panetta, K. 2021. 6 Key Takeaways from the Gartner Board of Directors Survey. *Gartner*. 21 October.

Paris 21. PARIS21 Launches Capacity Development 4.0 Guidelines.

Pauwels, E. 2020. The Anatomy of Information Disorders In Africa. Konrad Adenuer Foundation Office New York. 8 September.

Pentland, A. and H. Rahnama. 2022. The New Rules of Data Privacy. *Harvard Business Review*. 25 February.

Perri, L. 2022. What's New in Digital Government from the 2022 Gartner Hype Cycle. Gartner. 17 November.

Polley, L. 2022. To Disclose, or Not to Disclose, That Is the Question: A Methods-Based Approach for Examining & Improving the US Government's Vulnerabilities Equities Process. *RAND Corporation*. 11 March.

Ponemon Institute and SecureLlnk. 2021. *A Crisis in Third-Party Remote Access Security*. Austin, Texas: SecureLink, Inc.

Prevalent. 2022. 2022 Third-Party Risk Management Study.

PricewaterhouseCoopers. Disinformation Attacks Have Arrived in the Corporate Sector. Are You Ready?

Principles for Digital Development. How to Secure Private Data Stored and Accessed in the Cloud.

Privacy International. 10 Threats to Migrants and Refugees.

PRS Legislative Research. The Digital Personal Data Protection Bill, 2023.

Pruisenweg, W. 2021. Integrating Cyber Capacity into the Digital Development Agenda. The Hague: Global Forum on Cyber Expertise.

Ranking Digital Rights. Human Rights Impact Assessment: Samway Chapter Excerpt.

Ranking Digital Rights. The 2022 Ranking Digital Rights Big Tech Scorecard.

Ranking Digital Rights. Who We Are.

Rasner, G. 2022. *Cybersecurity and Third-Party Risk: Third Party Threat Hunting*. Hoboken, New Jersey: John Wiley and Sons.

RiskOptics (formerly Reciprocity). 2022. 10 Common Types of Digital Risks. *RiskOptics* (blog). 16 June.

RiskOptics (formerly Reciprocity). Complete Guide to NIST: Cybersecurity Framework.

RiskOptics (formerly Reciprocity). What Is Regulatory Compliance?

Rose, S. Planning for a Zero Trust Architecture: A Planning Guide for Federal Administrators. *National Institute of Standards and Technology White Paper*. No. 20.

Satariano, A. 2023. Europeans Take a Major Step Toward Regulating A.I. *The New York Times*. 14 June.

Secureframe. 2022. ISO 27000 Series: Wht the Standards are + their Purpose. 17 March.

Security Scorecard. 2020. How To Manage Third Party Digital Risk.

Shaak, E. 2022. PracticeMax Facing Class Action Over 2021 Data Breach Affecting 150K Patients. *ClassAction*. 28 July.

Shakarian, P. 2021. The Sunburst Hack Was Massive and Devastating – 5 Observations from a Cybersecurity Expert. *GovTech*. 4 January.

Shah, S. et. al. 2021. Reduce, Replace, Rethink: Transforming Technology Costs. *Bain & Co.* 10 July.

Shevlane, T. 2023. An Early Warning System for Novel AI Risks. *Google Deep Mind* (blog). 25 May.

Siemens. 2019. Siemens and Ponemon Institute study finds utility industry vulnerable to cyberattacks. Press release. 4 October.

Singh, A. and G. Triulzi. 2021. Technological Improvement Rate Predictions for All Technologies: Use of Patent Data and an Extended Domain Description. *ScienceDirect*. November.

Singtel. Climate Change and Environment.

Sinha, S. 2022. State of IoT 2022: Number of Connected IoT Devices Growing 18% to 14.4 Billion Globally.

Sissons, M. 2021. Our Commitment to Human Rights. *Meta*. 16 March.

Smith, S. 2022. Defending Ukraine: Early Lessons from the Cyber War. *Microsoft on the Issues* (blog). 22 June.

Spiekermann, S. 2022. What to Expect from IEEE 7000: The First Standard for Building Ethical Systems. *IEEE Technology and Society*. 21 February.

StartupDecisions. Complying with the Personal Data Act of Singapore.

State of California Office of the Attorney General. California Consumer Privacy Act.

Sullivan, M. 2023. What Is the Real Point of All These Letters Warning about AI? *Fast Company*. 31 May.

Sustainability Times. 2021. There's a Safer, Cleaner Way to Recover Rare-Earth Metals from Old Phones and Laptops. 23 April.

Sutherland, S. and S. Varley. 2022. Boards Must Work with CSOs on Value-Led Sustainability. *Ernst & Young*. 9 August.

Szalay, E. 2022. EU Should Ban Energy-Intensive Mode of Crypto Mining, Regulator Says. *Financial Times*. 19 January.

Tabansky, L. 2022. How Israel Became a Top Cyber Power. *The National Interest*. 21 March.

Tabassi, E. 2022. AI Risk Management Framework Initial Draft: 17 March 2022.

Tabassi, E. 2023. AI Risk Management Framework: AI RMF (1.0). Gaithersburg, Maryland: National Institute of Standards and Technology, 2023.

Tarchinski, M. 2022. Making the Business Case for Continuous Monitoring. *AuditBoard* (blog). February.

Malicious AI Report. The Malicious Use of Artificial Intelligence: Forecasting, Prevention, and Mitigation.

Target Internet. How Apple IOS14 Changes Have Affected Facebook Third-Party Tracking.

Tech Target. What Is a Data Breach?

Tech Target. What Is Federated Identity Management (FIM)? How Does It Work?

Terwangne, C. 2021. Council of Europe Convention 108+: A Modernised International Treaty for the Protection of Personal Data. *Computer Law & Security Review* 40: 105497.

*The Economist*. 2021. A Growing Number of Governments Are Spreading Disinformation Online. News release. 13 January.

The Center for Strategic and International Studies. 2022. Organizations and Nation-State Cyber Threats in the Crosshairs. 28 March.

The White House, Office of Science and Technology Policy. Blueprint for an AI Bill of Rights: Making Automated Systems Work for the American People.

The White House. 2021. National Security Memorandum on Improving Cybersecurity for Critical Infrastructure Control Systems. 28 July.

Think Tank European Parliament. European Declaration on Digital Rights and Principles.

Trilateral Research. 2022. A Survey of Artificial Intelligence Risk Assessment Methodologies: The global state of play and leading practices identified.

UK Information Commissioner's Office. Data Protection Self Assessment.

UK Information Commissioner's Office. Age appropriate design: a code of practice for online services.

UKCIS Digital Resilience Working Group. The Digital Resilience Framework.

UN Office of the Secretary-General's Envoy on Technology. Ensuring the Protection of Human Rights in the Digital Era.

UN Office of the Secretary-General's Envoy on Technology. Report of the Secretry-General: Roadmap for Digital Cooperation.

UN Digital Library. The age of digital interdependence: Report of the UN Secretary-General's High Level Panel on Digital Cooperation.

UN Development Programme. 2021. Fake News and Social Stability.

UN Development Programme. 2022. UNDP Tool to Fight Misinformation Scales Globally as a Digital Public Good. Press release. 17 May.

UN Environment Programme. 2021. The Growing Footprint of Digitalisation.

UN Institute for Training and Research. E-Waste Monitor.

UN High Commissioner for Human Rights. The right to privacy in the digital age–Report of the United Nations High Commissioner for Human Rights.

UNICEF. AI for Children.

University of Cambridge Centre for Alternative Finance. Cambridge Bitcoin Electricity Consumption Index.

United States Agency for International Development (USAID). Digital Ecosystem Framework.

US Equal Employment Opportunity Commission. 2021. EEOC Launches Initiative on Artificial Intelligence and Algorithmic Fairness. 28 October.

USAID. Cybersecurity Primer.

USAID. Digital Ecosystem Country Assessment (DECA) Toolkit: A How-To Guide for USAID Missions.

USAID. Digital Strategy.

VComply. 2020. What Is Business Continuity Risk?. 4 December.

Vera Solutions. 2019. 10 Criteria to Evaluate When Choosing a New Technology. *Vera Solutions* (blog). 17 October.

Wheeler, J. 2022. 2022: The Year of Digital Risk Discovery. *AuditBoard* (blog). 26 January.

Wheeler, J. 2020. 20 for 20: IRM Critical Capabilities and Top 20 Functions/Features. *Gartner* (blog). 20 October.

WHO. 2022. WHO Foresight: Monitoring Emerging Technologies and Building Futures-Thinking.

Wikipedia. Gartner Hype Cycle.

Williams, C. 2019. COSO ERM Framework – Background & Overview. *Strategic Decision Solutions* (blog). 11 March.

Withanage, S.V. and K. Habib. 2021. Life Cycle Assessment and Material Flow Analysis: Two Under-Utilized Tools for Informing E-Waste Management. *Sustainability*. 13 (14): 7939.

World Bank. Digital Government Readiness Assessment Toolkit: Guidelines for Task Teams.

World Bank. Digital Economy for Africa (DE4A)-Diagnostic Tool and Guidelines for Task Teams. Version 1.0.

World Bank. Digital Economy for Africa Country Diagnostic Tool and Guidelines for Task Teams. Version 2.0.

World Bank. GovTech Dataset.

World Bank. 2022. South Asia's Digital Opportunity: Accelerating Growth, Transforming Lives.

World Bank. World Development Report 2021: Data for Better Lives.

World Bank and United Nations. 2017. Combatting Cybercrime: Tools and Capacity Building for Emerging Economies. World Bank.

World Economic Forum. 17 Ways Technology Could Change the World by 2027.

World Economic Forum. Global Risks Report 2022.

World Economic Forum. 2019. Shaping the Future of Digital Economy and New Value Creation.

Wu, Y. 2022. AI in China: Regulations, Market Opportunities, Challenges for Investors. *China Briefing*. 14 October.

*Xinhua News Agency*. 2021. G20 Ministers Agree on Digital Working Group, Discuss Cybersecurity. 6 August.

Zurier, S. 2022. Code Vulnerability Failures in Manufacturing on Display in Toyota Supply Chain Attack. *SC Magazine*. 1 March.

www.ingramcontent.com/pod-product-compliance
Lightning Source LLC
Chambersburg PA
CBHW050243220326
41598CB00048B/7489